The Criminal Justice Student Writer's Manual

Second Edition

William A. Johnson, Jr.
Richard P. Rettig
Gregory M. Scott
Stephen M. Garrison

University of Central Oklahoma

Prentice
Hall

Upper Saddle River, New Jersey 07458

Library of Congress Cataloging-in-Publication Data

The criminal justice student writer's manual / William A. Johnson, Jr. . . . [et al.].—2nd ed.
    p.   cm.
  Includes bibliographical references and index.
  ISBN 0-13-093257-4
  1. Criminal justice, Administration of—United States—Authorship. 2. Criminal justice,
Administration of—Research—United States. 3. Legal composition—United States. 4.
Report writing—United States. I. Johnson, William A., 1942–

HV9950 .C74323 2002
808'.066364—dc21

2001033930

Publisher: Jeff Johnson
Executive Assistant: Brenda Rock
Senior Acquisitions Editor: Kim Davies
Assistant Editor: Sarah Holle
Managing Editor: Mary Carnis
Production Management: Clarinda
  Publication Services
Production Editor: Rosie Jones
Interior Design: Cindy Miller
Production Liaison: Adele M. Kupchik
Director of Manufacturing and Production:
  Bruce Johnson

Manufacturing Buyer: Cathleen Peterson
Senior Design Coordinator: Miguel Ortiz
Formatting: The Clarinda Company
Electronic Art Creation: The Clarinda Company
Marketing Manager: Ramona Sherman
Printer/Binder: R.R. Donnelley and Sons, Inc.
Copy Editor: Dave Prout
Proofreader: Nancy Hayes
Cover Design: Carey Davies
Cover Illustration: David Wink SIS/Images.Com
Cover Printer: Phoenix Color Corporation

Pearson Education LTD.
Pearson Education Australia PTY, Limited
Pearson Education Singapore, Pte. Ltd.
Pearson Education North Asia Ltd.
Pearson Education Canada, Ltd.
Pearson Educación de Mexico, S.A. de C.V.
Pearson Education—Japan
Pearson Education Malaysia, Pte. Ltd.

10 9 8 7 6 5 4 3
ISBN 0-13-093257-4

Contents

Preface　*vii*

Part One　*A Handbook of Style for Criminal Justice*　*1*

Chapter 1　**Writing as Communication**　**3**

Writing to Learn　3
   The Irony of Writing　3
   Challenge Yourself　5
   Maintain Self-Confidence　7
The Writing Process　7
   The Nature of the Process　7
   Finding a Thesis　10
   Defining Your Audience　13
   Invention Strategies　13
   Organizing Your Writing　16
   Drafting　18
   Revising　22
   Editing　23
   Proofreading　25

Chapter 2　**Writing Competently**　**26**

Guidelines for the Competent Writer　26
   Consider Your Audience　27
   Aim for Consistency　28
   Have Confidence in What You Already Know
     About Writing　28
   Eliminate Chronic Errors　28
Punctuation, Grammar, and Spelling　29
   Apostrophes　29
   Capitalization　30
   Colons　31
   Commas　32
   Dangling Modifiers　35
   Parallelism　36
   Fused (Run-on) Sentences　37
   Pronoun Errors　38
   Quotation Marks　40
   Semicolons　42

Sentence Fragments   43
Spelling   44

Chapter 3   **Formats   48**
Preliminary Considerations   48
Margins   49
Pagination   49
Title Page   50
Abstract   50
Executive Summary   51
Outline Summary   54
Table of Contents   55
Lists of Tables and Figures   56
Text   57
Headings and Subheadings   57
Tables   58
Illustrations and Figures   59
Reference Listing   59
Appendixes   59

Chapter 4   **Citing Sources   61**
Preliminary Decisions   61
What to Document   61
The Choice of Style   61
Citing Sources in ASA Style   62
Text Citations in ASA Style   62
References in ASA Style   70
Citing Sources in APA Style   83
Text Citations in APA Style   83
References in APA Style   89

Part Two   *How to Conduct Research in Criminal Justice*   **99**

Chapter 5   **Organizing the Research Process   101**
Gaining Control   101
Who Is in Control of Your Paper   101
Understand Your Assignment   102
What Is Your Topic   102
What Is Your Purpose   102
Who Is Your Audience   102
What Kind of Research Are You Doing   103
Keep Your Perspective   104
Effective Research Methods   104
Establish an Effective Procedure   104
Give Yourself Plenty of Time   105
Do Background Reading   105
Narrow Your Topic and Establish a Working Thesis   106
Develop a Working Bibliography   107

Write for Needed Information   107
Evaluate Written Sources   108
Use Photocopies and Download Articles   109
Determine Whether to Conduct Interviews or Surveys   109
Draft a Thesis and Outline   110
Write a First Draft   110
Obtain Feedback   110
Ethical Use of Source Material   111
When to Quote   111
Paraphrasing   112
Avoiding Plagiarism   112

Chapter 6   Information Sources and Distance Learning   114
Library Research   114
Criminal Justice on the World Wide Web   117
A Guide to Distance Learning   120
For Students Considering Distance Learning   120
Criminal Justice Distance Learning Courses and Distance Learning
Resources On-line   123
For Students About to Take or Taking On-line Courses   123

Part Three   *Writing Different Types of Criminal Justice Papers*   125

Chapter 7   Brief Writing Assignments   127
Reaction Papers   127
Select a Suitable Reaction Statement   127
Explain Your Selection   128
Clearly Define the Issue Addressed in the Statement   128
Clearly State Your Position on the Issue   129
Defend Your Position   129
Conclude Concisely   129
Article Critiques   129
Choosing an Article   132
Writing the Critique   132
Book Reviews   134
Objectives of a Book Review   134
Elements of a Book Review   141
Types of Book Reviews: Reflective and Analytical   142
Format and Length of a Book Review   143

Chapter 8   Criminal Justice Agency Case Studies   144
What Is a Case Study   144
Using Case Studies in Research   145
Limitations of the Case Study Method   145
Types of Case Studies Written in Criminal Justice   146
How to Conduct a Criminal Justice Agency Case Study   146
Selecting a Topic   147
The Importance of Interviews   148

Elements of the Case Study Paper   148
Overview of Contents   148
Text   149
Summary   150

Chapter 9   Criminal Justice Policy Analysis Papers   152
What Is Policy Analysis   152
Policy Analysis Research Proposals   154
Introduction to Policy Analysis Research Proposals   154
The Purpose of Research Proposals   154
The Content of Research Proposals   155
Criminal Justice Policy Analysis Papers   161
Definition of a Policy Analysis Paper   161
Purpose of Policy Analysis Papers   161
Contents of a Criminal Justice Policy Analysis Paper   162
Format of a Policy Analysis Paper   162
Presentation of Policy Analysis Papers   171

Glossary   172
References   208
Index   210

Preface

The more complex criminal justice systems become, the greater the need for clear, direct communication. The sad reality is that many people have difficulty writing a simple declarative sentence. In all areas of criminal justice, no skill is recognized as more important than the ability to get messages on paper clearly in order to get business done. If you learn to write well, you will be more valuable in whatever line of work you pursue.

Administrators in criminal justice agencies are looking for people who can condense a mass of data to one sheet of clear comment. They are looking for men and women who can report the day-to-day activities of a complex system concisely and accurately. Administrators want more than undigested, descriptive data; they want to know what the experts in each area believe the data mean.

This book is designed to help you improve your writing. You will find principles and guidelines to help you write complex reports as well as summarize information into condensed presentations. Good writers know and use the terminology of their discipline. In writing reports and professional communication, criminal justice professionals often need to draw from the lexicon of several disciplines, e.g. law enforcement, forensics, psychology, sociology, law, and management. We have included a broad and extensive *glossary* to aid you in writing the assignments in this manual. As you continue to use this manual throughout your college and professional career, we hope your writing skills evolve and aid you in attaining your professional goals and objectives.

ACKNOWLEDGMENTS

We would like to thank Kim Davies, Sarah Holle, Miguel Ortiz, Mary Carnis, Rosie Jones, and all those who assisted in compiling and editing this book. We would also like to thank the following reviewers for their helpful comments:

Robert Meadows, Californian Lutheran University, Thousand Oaks, Ca.
Ronald Holmes, University of Louisville, Louisville, Ky.
Susann Welty-Barr, University of Arkansas, Little Rock, Ak.
Janet Hageman, San Jose State University, San Jose, Ca.

William Johnson
Richard Rettig
Greg Scott
Stephen Garrison

To
Chris

A Handbook of Style
for Criminal Justice

Chapter 1 Writing as Communication

Chapter 2 Writing Competently

Chapter 3 Formats

Chapter 4 Citing Sources

Writing as Communication

WRITING TO LEARN

Writing is a way of ordering your experience. Think about it: No matter what you are writing—it may be a paper for your introductory criminal justice class, a short story, a limerick, a grocery list—you are putting pieces of your world together in new ways and making yourself freshly conscious of these pieces. This is one of the reasons writing is so hard. From the infinite welter of data that your mind continually processes and locks in your memory, you are selecting only certain items significant to the task at hand, relating them to other items, and phrasing them in a new coherence. You are mapping a part of your universe that has hitherto been unknown territory. You are gaining a little more control over the processes by which you interact with the world around you.

Writing is, therefore, one of the best ways to learn. This statement may sound odd at first. If you are an unpracticed writer, you may share a common notion that the only purpose writing can have is to express what you already know or think. Any learning that you as a writer might do has already been accomplished by the time your pen meets the paper. In this view, your task is to inform and even surprise the reader. But if you are a practiced writer, you know that at any moment as you write, you are capable of surprising yourself by discovering information you never knew that you knew. And it is surprise that you look for: the shock of seeing what happens in your own mind when you drop an old, established opinion into a batch of new facts or bump into a cherished belief from a different angle. Writing synthesizes new understanding for the writer. E. M. Forster's famous question—"How do I know what I think until I see what I say?"—is one that all of us could ask. We make meaning as we write, jolting ourselves by little discoveries into a larger and more interesting universe.

The Irony of Writing

Good writing helps the reader become aware of the ironies and paradoxes of human existence. One such paradox is that good writing expresses both that which is unique about the writer and, at the same time, that which is common, not to the writer alone, but to every human being. Many of our most famous political statements share this double attribute of mirroring the singular and the ordinary. For example, read the following excerpts from President Franklin Roosevelt's first inaugural address, spoken on March 4, 1933, in the middle of the Great Depression, then answer this question: Is what Roosevelt says famous in history because its expression is extraordinary or because it appeals to something that is basic to every human being?

> *This is pre-eminently the time to speak the truth, the whole truth, frankly and boldly. Nor need we shrink from honestly facing conditions in our country today. This great nation will endure as it has endured, will revive and will prosper.*
>
> *So first of all let me assert my firm belief that the only thing we have to fear is fear itself—nameless, unreasoning, unjustified terror which paralyzes needed efforts to convert retreat into advance.*
>
> *In every dark hour of our national life a leadership of frankness and vigor has met with that understanding and support of the people themselves which is essential to victory. I am convinced that you will again give that support to leadership in these critical days.*
>
> *In such a spirit on my part and on yours we face our common difficulties. They concern, thank God, only material things. Values have shrunken to fantastic levels; taxes have risen; our ability to pay has fallen, government of all kinds is faced by serious curtailment of income; the means of exchange are frozen in the currents of trade; the withered leaves of industrial enterprise lie on every side; farmers find no markets for their produce; the savings of many years in thousands of families are gone.*
>
> *More important, a host of unemployed citizens face the grim problem of existence, and an equally great number toil with little return. Only a foolish optimist can deny the dark realities of the moment.*
>
> *Yet our distress comes from no failure of substance. We are stricken by no plague of locusts. Compared with the perils which our forefathers conquered because they believed and were not afraid, we have still much to be thankful for. Nature still offers her bounty and human efforts have multiplied it. Plenty is at our doorstep, but a generous use of it languishes in the very sight of the supply. . . .*
>
> *The measure of the restoration lies in the extent to which we apply social values more noble than mere monetary profit.*
>
> *Happiness lies not in the mere possession of money; it lies in the joy of achievement, in the thrill of creative effort.*
>
> *The joy and moral stimulation of work no longer must be forgotten in the mad chase of evanescent profits. These dark days will be worth all they cost us if they teach us that our true destiny is not to be ministered unto but to minister to ourselves and to our fellow-men. (Commager 1963:240)*

The help that writing gives us with learning and with controlling what we learn is one of the major reasons why your criminal justice instructors will require a great deal of writing from you. Learning the complex and diverse world of the criminal justice professional takes more than a passive ingestion of facts. You have to come to grips with social issues and with your own attitudes toward them. When you write in a class on criminal justice or juvenile delinquency, you are entering into the world of professional researchers in the same way they do—testing theory against fact and fact against belief.

> ## Learning by Writing
>
> A way of testing the notion that writing is a powerful learning tool is by rewriting your notes from a recent class lecture. The type of class does not matter; it can be history, chemistry, criminal justice, whatever. If possible, choose a difficult class, one in which you are feeling somewhat unsure of the material and one for which you have taken copious notes.
>
> As you rewrite, provide the *transitional elements* (the connecting phrases like *in order to, because of, and, but, however*) that you were unable to supply in class because of the press of time. Furnish your own examples or illustrations of the ideas expressed in the lecture.
>
> This experiment will force you to supply necessary coherence out of your own thought processes. See if the loss of time it takes you to rewrite the notes is not more than compensated for by a gain in your understanding of the lecture material.

Virtually everything that happens in the discipline of criminal justice happens on paper first. Documents are wrestled into shape before their contents can affect the public. Meaningful social programs are written before they are implemented. The written word has helped bring slaves to freedom, end wars, create new opportunities in the workplace, and shape the values of nations. Often, gaining recognition for ourselves and our ideas depends less upon what we say than upon how we say it. Accurate and persuasive writing is absolutely vital to the criminal justice professional.

Challenge Yourself

There is no way around it: writing is a struggle. Do you think you are the only one to feel this way? Take heart! Writing is hard for everybody, great writers included. Bringing order into the world is never easy. Isaac Bashevis Singer, winner of the 1978 Nobel Prize in literature, once wrote, "I believe in miracles in every area of life except writing. Experience has shown me that there are no miracles in writing. The only thing that produces good writing is hard work" (Lunsford and Connors 1992:2).

Hard work was evident in the words of John F. Kennedy's Inaugural Address. As you read the following excerpts from Kennedy's speech, what images come to mind? Historians tend to consider a president "great" when his words live longer than his deeds in the minds of the people. Do you think this will be—or has been—true of Kennedy?

> *We observe today not a victory of party but a celebration of freedom—symbolizing an end as well as a beginning—signifying renewal as well as change. For I have sworn before you and Almighty God the same solemn oath our forebears prescribed nearly a century and three-quarters ago.*
>
> *The world is very different now. For man holds in his mortal hands the power to abolish all forms of human poverty and all forms of human life. And*

yet the same revolutionary beliefs for which our forebears fought are still at issue around the globe—the belief that the rights of man come not from the generosity of the state but from the hand of God.

We dare not forget today that we are the heirs of that first revolution. Let the word go forth from this time and place, to friend and foe alike, that the torch has been passed to a new generation of Americans—born in this century, tempered by war, disciplined by a hard and bitter peace, proud of our ancient heritage—and unwilling to witness or permit the slow undoing of those human rights to which this nation has always been committed, and to which we are committed today at home and around the world. . . .

In the long history of the world, only a few generations have been granted the role of defending freedom in its hours of maximum danger. I do not shrink from this responsibility—I welcome it. I do not believe that any of us would exchange places with any other people or any other generation. The energy, the faith, the devotion which we bring to this endeavor will light our country and all who serve it—and the glow from that fire can truly light the world.

And so, my fellow Americans: ask not what your country can do for you— ask what you can do for your country.

My fellow citizens of the world: ask not what America will do for you, but what together we can do for the freedom of man. (Commager 1963:688–689)

One reason writing is difficult is that it is not actually a single activity at all but a process consisting of several activities that can overlap each other, with two or more sometimes operating *simultaneously* as you labor to organize and phrase your thoughts. (We will discuss these activities later in this chapter.) The writing process tends to be sloppy for everyone; it is often a frustrating search for meaning and for the best way to articulate that meaning.

Frustrating though that search may sometimes be, it need not be futile. Remember this: The writing process makes use of skills that we all have. The ability to write, in other words, is not some magical competence bestowed on the rare, fortunate individual. While few of us may achieve the proficiency of Isaac Singer, we are all capable of phrasing thoughts clearly and in a well-organized fashion. But learning how to do so takes *practice.*

The one sure way to improve your writing is to write. One of the toughest but most important jobs in writing is to maintain enthusiasm for your writing project. Commitment may sometimes be complicated, given the difficulties that are inherent in the writing process and that can be made worse when the project is unappealing at first glance. How, for example, can you be enthusiastic about having to write a paper analyzing prison reform when you know little about the American correctional system and see no real use in writing about it?

One of the worst mistakes that unpracticed student writers make is to fail to assume responsibility for keeping themselves interested in their writing. No matter how hard it may seem at first to drum up interest in your topic, *you have to do it*—that is, if you want to write a paper you can be proud of, one that contributes useful material and a fresh point of view to the topic. One thing is guaranteed: If you are bored with your writing, your reader will be, too. So what can you do to keep your interest and energy level high?

Challenge yourself. Think of the paper not as an assignment but as a piece of writing that has a point to make. To get this point across persuasively is the real reason why you are writing, *not* the simple fact that a teacher has assigned you a project. If someone were to ask you why you are writing your paper and your immediate, unthinking response is, "Because I've been given a writing assignment," or "Because I want a good grade," or some other nonanswer along these lines, your paper may be in trouble.

If, on the other hand, your first impulse is to explain the challenge of your main point—"I'm writing to show how prison reform will benefit both inmates and the American taxpayer"—then you are thinking usefully about your topic.

Maintain Self-Confidence

Having confidence in your ability to write well about your topic is essential for good writing. This does not mean that you will always know what the end result of a particular writing activity will be. In fact, you have to cultivate your ability to tolerate a high degree of uncertainty while weighing evidence, testing hypotheses, and experimenting with organizational strategies and wording. Be ready for temporary confusion and for seeming dead ends, and remember that every writer faces these obstacles. It is from your struggle to combine fact with fact, to buttress conjecture with evidence, that order arises.

Do not be intimidated by the amount and quality of work that others have already done in your field of inquiry. The array of opinion and evidence that confronts you in the published literature can be confusing. But remember that no important topic is ever exhausted. *There are always gaps*—questions that have not yet been satisfactorily explored either in the published research on a subject or in the prevailing popular opinion. It is in these gaps that you establish your own authority, your own sense of control.

Remember that the various stages of the writing process reinforce each other. Establishing a solid motivation strengthens your sense of confidence about the project, which in turn influences how successfully you organize and write. If you start out well, using good work habits, and give yourself ample time for the various activities to coalesce, you should produce a paper that will reflect your best work, one that your audience will find both readable and useful.

THE WRITING PROCESS

The Nature of the Process

As you engage in the writing process, you are doing many different things at once. While planning, you are no doubt defining the audience for your paper at the same time that you are thinking about the paper's purpose. As you draft the paper, you may organize your next sentence while revising the one you have just written. Different parts of the writing process overlap, and much of the difficulty of writing is that so many things happen at once. Through practice—in other words, through *writing*—it is possible to learn how to control those parts of the process that can in fact be controlled and to encourage those mysterious, less controllable activities.

No two people go about writing in exactly the same way. It is important for you to recognize routines—modes of thought as well as individual exercises—that help you negotiate the process successfully. And it is also important to give yourself as much time as possible to complete the process. Procrastination is one of the writer's greatest enemies. It saps confidence, undermines energy, destroys concentration. Working regularly and following a well-planned schedule as closely as possible often make the difference between a successful paper and an embarrassment.

Although the various parts of the writing process are interwoven, there is naturally a general order to the work of writing. You have to start somewhere! What follows is a description of the various stages of the writing process—*planning, drafting, revising, editing, and proofreading*—along with suggestions on how to approach each most successfully.

PLANNING

Planning includes all activities that lead up to the writing of the first draft. The particular activities in this stage differ from person to person. Some writers, for instance, prefer to compile a formal outline before writing that draft. Some writers perform brief writing exercises to jump-start their imaginations. Some draw diagrams; some doodle. Later we will look at a few starting strategies, and you can determine which may help you.

Now, however, let us discuss certain early choices that all writers must make during the planning stage. These choices concern *topic, purpose,* and *audience,* three elements that make up the *writing context,* or the terms under which we all write. Every time you write, even if you are writing a diary entry or a note to the milkman, these elements are present. You may not give conscious consideration to all of them in each piece of writing that you do, but it is extremely important to think carefully about them when writing a criminal justice paper. Some or all of these defining elements may be dictated by your assignment, yet you will always have a degree of control over them.

SELECTING A TOPIC

No matter how restrictive an assignment may seem, there is no reason to feel trapped by it. Within any assigned subject you can find a range of topics to explore. What you are looking for is a topic that engages your own interest. Let your curiosity be your guide. If, for example, you have been assigned the subject of prison reform, then guide yourself to find some issue concerning prison reform that interests you. (How does inmate education—for example, taking college courses by correspondence or TV—affect recidivism? Should inmates incarcerated for violent offenses be allowed to "bulk up" by lifting weights?) Any good topic comes with a set of questions; you may well find that your interest picks up if you simply begin asking questions.

One strong recommendation: Ask your questions *on paper.* Like most other mental activities, the process of exploring your way through a topic is transformed when you write down your thoughts as they come instead of letting them fly through your mind unrecorded. Remember the old adage from Louis Agassiz: "A pen is often the best of eyes" (Pearce 1958:106).

While it is vital to be interested in your topic, you do not have to know much about it at the outset of your investigation. In fact, having too heartfelt a commitment to a topic can be an impediment to writing about it; emotions can get in the way of objectivity.

Better often to choose a topic that has piqued your interest yet remained something of a mystery to you: a topic discussed in one of your classes, perhaps, or mentioned on television or in a conversation with friends.

NARROWING A TOPIC

The task of narrowing your topic offers you a tremendous opportunity to establish a measure of control over the writing project. It is up to you to hone your topic to just the right shape and size to suit both your own interests and the requirements of the assignment. Do a good job of it, and you will go a long way toward guaranteeing yourself sufficient motivation and confidence for the tasks ahead of you. If you do not narrow your topic well, somewhere along the way you may find yourself directionless and out of energy.

Generally, the first topics that come to your mind will be too large to handle in your research paper. For example, the topic of gun control has generated a tremendous number of published news articles and reports recently by experts in the field. Despite all the attention turned toward this topic, however, there is still plenty of room for you to investigate it on a level that has real meaning to you and that does not merely recapitulate the published research. What about an analysis of how enactment of the Brady bill has affected handgun crimes in your city? The problem with most topics is not that they are too narrow or too completely explored, but rather that they are so rich it is difficult to choose the most useful ways to address them. Take some time to narrow your topic. Think through the possibilities that occur to you, and as always, jot down your thoughts.

Narrowing Topics

Without doing research, see how you can narrow the following general topics; for example:

- *General topic:*   Juvenile delinquency
- *Narrowed topics:*   Labeling serious habitual offenders: harassment or public safety?

　　　　　　　　　　　　 Substance abuse and delinquent behavior

　　　　　　　　　　　　 Kids that kill: juvenile delinquents or adult murderers?

GENERAL TOPICS

Crime in America	Political corruption
International terrorism	Costs of incarceration
Education	Affirmative action hiring policies
Freedom of speech	Freedom to bear arms
Gun control	Training police officers

Students in an undergraduate course on criminology were told to write an essay of 2500 words on one of the following topics. Next to each general topic is an example of how students narrowed it to make a manageable paper topic.

General Topic	*Paper Topic*
Homicide	The effect of homicide on the black male population in the United States
Teenage Crimes	The role of drug involvement in teenage crimes in the United States
Prisons	Should U.S. prisons be run by the private sector?

Finding a Thesis

As you plan your writing, be on the lookout for an idea that can serve as your *thesis*. A thesis is not a fact, which can be immediately verified by data, but an assertion worth discussing, an argument with more than one possible conclusion. Your thesis sentence will reveal to your reader not only the argument you have chosen but also your orientation toward it and the conclusion that your paper will attempt to prove.

In looking for a thesis, you are doing many jobs at once:

1. You are limiting the amount and kind of material that you must cover, thus making it manageable.
2. You are increasing your own interest in the narrowing field of study.
3. You are working to establish your paper's purpose, the *reason* why you are writing about your topic. (If the only reason you can see for writing is to earn a good grade, then you probably won't!)
4. You are establishing your notion of who your audience is and what sort of approach to the subject might best catch their interest.

In short, you are gaining control over your writing context. For this reason, it is a good idea to come up with a thesis early on, a *working thesis,* which will very probably change as your thinking deepens but which will allow you to establish a measure of order in the planning stage.

THE THESIS SENTENCE

The introduction of your paper will contain a sentence that expresses the task that you intend to accomplish. This thesis sentence communicates your main idea, the one you are going to support, defend, or illustrate. The thesis sets up an expectation in the reader's mind that is your job to satisfy. But in the planning stage a thesis sentence is more than just the statement that informs your reader of your goal. It is a valuable tool to help you narrow your focus and confirm in your own mind your paper's purpose.

DEVELOPING A THESIS

A Crime and Society class was assigned to write a twenty-page paper studying a problem currently being faced by the municipal authorities in their own city. The choice of the problem was left up to the students. One class member, Mark Gonzales, decided to investigate the problem posed to the city by the large number of abandoned buildings in a downtown neighborhood that he drove through on his way to the university. His first working thesis read as follows:

`Abandoned houses breed crime.`

The problem with this thesis, as Mark found out, was that it was not an idea that could be argued but rather a fact that could be easily corroborated by the sources he began to consult. As Mark read reports from such sources as the Urban Land Institute and the City Planning Commission and talked with representatives from the Community Planning Department, he began to get interested in the dilemma his city faced in responding to the problem of abandoned buildings. Here is Mark's second working thesis:

`Removal of abandoned buildings is a major problem facing the city.`

This thesis narrowed the topic somewhat and gave Mark an opportunity to use material from his research, but there was still no real comment attached to it. It still stated a bare fact, easily proved. At this point, Mark became interested in the still narrower topic of how building removal should best be handled. He found that the major issue was funding the demolition and that different civic groups favored different funding methods. As Mark explored the arguments for and against funding plans, he began to feel that one of them might be best for the city. Mark's third working thesis:

`Providing alternative shelter for homeless people reduces crime associated with abandoned buildings.`

Note how this thesis narrows the focus of Mark's paper even further than the other two while also presenting an arguable hypothesis. It tells Mark what he has to do in his paper, just as it tells his reader what to expect.

At some time during your preliminary thinking on a topic, you should consult the library to see how much published work has already been done. This search is beneficial in at least two ways:

1. It acquaints you with a body of writing that will become very important in the research phase of the paper.
2. It gives you a sense of how your topic is generally addressed by the community of scholars you are joining. Is the topic as important as you think it is? Has there already been so much research on the topic as to make your inquiry, in its present formulation, irrelevant?

As you go about determining your topic, remember that one goal of criminal justice writing in college is to enhance your own understanding of the social and social-psychological process, to build an accurate model of the way social institutions work. Let this goal help you to aim your research into those areas that you know are important to your knowledge of the discipline.

DEFINING A PURPOSE

There are many ways to classify the purposes of writing, but in general, most writing is undertaken either to inform or to persuade an audience. The goal of informative or expository writing is, simply, to impart information about a particular subject, while the aim of persuasive writing is to convince your reader of your point of view on an issue. The distinction between expository and persuasive writing is not hard and fast, and most criminal justice writing has elements of both types. Most effective writing, however, is clearly focused on either exposition or persuasion. When you begin writing, consciously select a primary approach of exposition or persuasion and then set out to achieve that goal.

Suppose you have been required to write a paper explaining how parents' attitudes affect their children's choice of colleges. If you are writing an expository paper, your task could be to describe in as coherent and impartial a way as possible the attitudes of the parents and the choices of their children.

If, however, your paper attempts to convince your reader that parental attitudes often result in children making poor choices, you are now writing to persuade, and your strategy is radically different. You will now need to explain the negative effects of parental attitudes. Persuasive writing seeks to influence the opinions of its audience toward its subject.

Know what you want to say. By the time of your final draft, you must have a very sound notion of the point you wish to argue. If, as you write that final draft, someone were to ask you to state your thesis, you should be able to give a satisfactory answer with a minimum of delay and no prompting. If, on the other hand, you have to hedge your answer because you cannot easily express your thesis, you may not yet be ready to write a final draft.

Watch out for bias! There is no such thing as pure objectivity. You are not a machine. No matter how hard you may try to produce an objective paper, the fact is that every choice you make as you write is influenced to some extent by your personal beliefs and opinions. What you tell your readers is *truth*, in other words, is influenced, often without your knowing, by a multitude of factors: your environment, upbringing, and education; your attitude toward your audience; your political affiliation; your race and gender; your career goals; and your ambitions for the paper you are writing. The influence of such factors can be very subtle, and it is something you must work to identify in your own writing as well as in the writing of others in order not to mislead

Knowing What You Want to Say

Two writers have been asked to state the theses of their papers. Which one of the writers better understands the writing task?

- *Writer 1:* "My paper is about police-community relations."
- *Writer 2:* "My paper argues that improving communication between the police and citizens in the community raises morale among police officers and helps people take greater responsibility for important issues within the community."

The second writer has a clear view of her task. The first knows what her topic is—police-community relations—but may not yet know what it is about these relations that fosters important changes within the community. It may be that you will have to write a rough draft or two or engage in various prewriting activities in order to arrive at a secure understanding of your task.

or be misled. Remember that one of the reasons for writing is *self-discovery.* The writing you will do in criminal justice classes—as well as the writing you will do for the rest of your life—will give you a chance to discover and confront honestly your own views on your subjects. Responsible writers keep an eye on their own biases and are honest about them with their readers.

Defining Your Audience

In any class that requires you to write, it may sometimes be difficult to remember that the point of your writing is not simply to jump through the technical hoops imposed by the assignment. The point is *communication,* the transmission of your knowledge and your conclusions to the reader in a way that suits you. Your task is to pass to your reader the spark of your own enthusiasm for your topic. Readers who were indifferent to your topic should look at it in a new way after reading your paper. This is the great challenge of writing: to enter into your reader's mind and leave behind both new knowledge and new questions.

It is tempting to think that most writing problems would be solved if the writer could view the writing as if it had been produced by another person. The discrepancy between the understanding of the writer and that of the audience is the single greatest impediment to accurate communication. To overcome this barrier, you must consider your audience's needs. By the time you begin drafting, most if not all of your ideas will have begun to attain coherent shape in your mind, so that virtually any words in which you try to phrase those ideas will reflect your thought accurately—*to you.* Your readers, however, do not already hold the conclusions that you have so painstakingly achieved. If you omit from your writing the material that is necessary to complete your readers' understanding of your argument, they may well not be able to supply that information themselves.

The potential for misunderstanding is present for any audience, whether it is made up of general readers, experts in the field, or your professor, who is reading, in part, to see how well you have mastered the constraints that govern the relationship between writer and reader. Make your presentation as complete as possible, bearing in mind your audience's knowledge of your topic.

Invention Strategies

We have discussed various methods of selecting and narrowing the topic of a paper. As your focus on a specific topic sharpens, you will naturally begin to think about the kinds of information that will go into the paper. In the case of papers not requiring formal research, that material comes largely from your own recollections. Indeed, one of the reasons instructors assign such papers is to convince you of the incredible richness of your memory, the vastness and variety of the "database" you have accumulated and which, moment by moment, you continue to build.

So vast is your horde of information that it is sometimes difficult to find within it the material that would best suit your paper. In other words, finding out what you already know about a topic is not always easy. *Invention,* a term borrowed from classical rhetoric, refers to the task of discovering, or recovering from memory, such information. As we write, we go through some sort of invention procedure that helps us explore our topic. Some writers seem to have little problem coming up with material; others need more help. Over the centuries writers have devised different exercises that can help locate useful material housed in memory. We shall look at a few of these briefly.

FREEWRITING

Freewriting is an activity that forces you to get something down on paper. There is no waiting around for inspiration. Instead, you set yourself a time limit—perhaps three or five minutes—and write for that length of time without stopping, not even to lift the pen from the paper or your hands from the keyboard. Focus on the topic, and don't let the difficulty of finding relevant material stop you from writing. If necessary, you may begin by writing, over and over, some seemingly useless phrase, such as "I cannot think of anything to write about" or perhaps the name of your topic. Eventually, something else will occur to you. (It is surprising how long a three-minute freewriting session can seem to last!) At the end of the freewriting, look over what you have produced for anything of use. Much of the writing will be unusable, but there may be an insight or two that you did not know you possessed.

Besides helping to recover from your memory usable material for your paper, freewriting has other benefits. First, it takes little time to do, which means you may repeat the exercise as often as you like within a relatively short span of time. Second, it breaks down some of the resistance that stands between you and the act of writing. There is no initial struggle to find something to say; you just *write*.

Freewriting

The teacher in Shelby Johnson's second-year Family as a Social Institution class assigned Shelby a paper to write focusing on some aspect of American family life. Shelby, who felt her understanding of the family as an institution was slight, tried to get her mind started on the job of finding a topic that interested her with a two-minute freewriting. Thinking about the family and child development, Shelby wrote steadily for three minutes without lifting her pen from the paper. Here is the result of her freewriting:

> Family family family family family family family what do I know? My family— father mother sister. Joely. Parents Mom and Dad. Carole and Don. Child development. My development. Okay okay okay okay okay. Both parents were present all my life. Both worked. Professionals. Dad at the school, Mom at the office. Sometimes we wished Mom was at home. The bedtimes she missed, Joely griping and that night she cried. That old empty feeling. Emptinessssssss. The way the air conditioner sounded when the house was quiet. That might be interesting: working parents, the effects on kids. A personal view. Two-paycheck families. Necessary nowadays—why not before? What's happened to make two jobs necessary. No. Back up. Go back to life in two-income family. I like it. Where to start. I could interview Mom. Get recent statistics on two-paycheck families. Where???? Ask in library tomorrow.

BRAINSTORMING

Brainstorming is simply making a list of ideas about a topic. It can be done quickly and at first without any need to order items into a coherent pattern. The point is to write down everything that occurs to you quickly and as briefly as possible, using individual words or short phrases. Once you have a good-sized list of items, you can then group the items according to relationships that you see among them. Brainstorming thus allows you to uncover both ideas stored in your memory and useful associations among those ideas.

Brainstorming

A professor in a criminal justice class asked her students to write a 700-word paper, in the form of a letter to be translated and published in a Warsaw newspaper, giving Polish readers useful advice about living in a democracy. One student, Chelsea Blake, started thinking about the assignment by brainstorming. First, she simply wrote down anything that occurred to her:

Voting rights	Welfare	Freedom of press
Protest movements	Everybody equal	Minorities
Racial prejudice	The American Dream	Injustice
The individual	No job security	Lobbyists and PACs
Justice takes time	Psychological factors	Aristocracy of wealth
Size of bureaucracy	Market economy	Many choices

Thinking through her list, Chelsea decided to rearrange it into two lists, one devoted to positive aspects of life in a democracy, the other to negative aspects. At this point she decided to discard some items that were redundant or did not seem to have much potential. As you can see, Chelsea had some questions about where some of her items would fit.

POSITIVE	NEGATIVE
Voting rights	Aristocracy of wealth
Freedom of the press	Justice takes time
Everybody equal	Racial prejudice
The American Dream	Welfare
Psychological factors	Lobbyists and PACs
Protest movements (positive?)	Size of bureaucracy

At this point, Chelsea decided that her topic would be the ways in which money and special interests affect a democratically elected government. Which items on her lists would be relevant to Chelsea's paper?

Asking Questions

For a class in Criminal Law, a professor asked her class to write a paper describing the impact of Supreme Court clerks on the decision-making process. One student developed the following questions as he began to think about a thesis:

- *Who* are the Supreme Court's clerks? (How old? Of what ethnicity and gender are they? What are their politics?)
- *What* are their qualifications for the job?
- *What* exactly is their job?
- *When* during the court term are they most influential?
- *Where* do they come from? (Is there any geographical pattern discernible in the way they are chosen? Any pattern regarding religion? Do certain law schools contribute a significantly greater number of clerks than any others?)
- *How* are they chosen? (appointed? elected?)
- *When* in their careers do they serve?
- *Why* are they chosen as they are?
- *Who* have been some influential court clerks? (Have any gone on to sit on the bench themselves?)

Can you think of other questions that would make for useful inquiry?

ASKING QUESTIONS

It is always possible to ask most or all of the following questions about any topic: *Who? What? When? Where? Why? How?* These questions force you to approach the topic as a journalist does, setting it within different perspectives that can then be compared.

MAINTAINING FLEXIBILITY

As you engage in invention strategies you are also doing other work. You are still narrowing your topic, for example, as well as making decisions that will affect your choice of tone or audience. You are moving forward on all fronts, with each decision you make affecting the others. This means you must be flexible to allow for slight adjustments in your understanding of the paper's development and of your goal. Never be so determined to prove a particular theory that you fail to notice when your own understanding of it changes. *Stay objective.*

Organizing Your Writing

A paper that has all the facts but gives them to the reader in an ineffective order will confuse rather than inform or persuade. While there are various methods of grouping ideas, none is potentially more effective than *outlining*. Unfortunately, no organizing process is more often misunderstood.

OUTLINING FOR YOURSELF

Outlining can do two jobs. First, it can force you, the writer, to gain a better understanding of your ideas by arranging them according to their interrelationships. There is one primary rule of outlining: Ideas of equal weight are placed on the same level within the outline. This rule requires you to determine the relative importance of your ideas. You have to decide which ideas are of the same type or order and into which subtopic each idea best fits.

If, in the planning stage, you carefully arrange your ideas in a coherent outline, your grasp of the topic will be greatly enhanced. You will have linked your ideas logically together and given a basic structure to the body of the paper. This sort of subordinating and coordinating activity is difficult, however, and as a result, inexperienced writers sometimes begin to write their first draft without an effective outline, hoping for the best. This hope is usually unfulfilled, especially in complex papers involving research.

OUTLINING FOR YOUR READER

The second job an outline can perform is to serve as a reader's blueprint to the paper, summarizing its points and their interrelationships. A busy person can quickly get a sense of your paper's goal and the argument you have used to promote it by consulting your outline. The clarity and coherence of the outline helps to determine how much attention your audience will give to your ideas.

While neither the American Psychological Association (APA) nor the American Sociological Association (ASA) in their style guides formally require the inclusion of an outline with a paper submitted for publication to a professional journal, both the *Publication Manual of the American Psychological Association* (1994:90–93) and the

Organizing Thoughts

Juan, a student in a second-year criminal justice class, researched the impact of a worker-retraining program in his state and came up with the following facts and theories. Number them in logical order.

_____ A growing number of workers in the state do not possess the basic skills and education demanded by employers.

_____ The number of dislocated workers in the state increased from 21,000 in 1982 to 32,000 in 1992.

_____ A public policy to retrain uneducated workers would allow them to move into new and expanding sectors of the state economy.

_____ Investment in high technology would allow the state's employers to remain competitive in the production of goods and services in both domestic and foreign markets.

_____ The economy is becoming more global and more competitive.

American Sociological Association Style Guide (1997:19–20) advocate the use of organizational headings, based on formal outline patterning, within the paper's text. Indeed, a formal outline is such a useful tool that your criminal justice instructor may require you to submit one. A look at the model presented in other chapters of this manual will show you how strictly these formal outlines are structured. But while you must pay close attention to the requirements of the accompanying outline, do not forget how powerful a tool an outline can be in the early planning stages of your paper.

FORMAL OUTLINE PATTERN

Following this pattern accurately during the planning stage of your paper helps to guarantee that your ideas are placed logically:

Thesis sentence (prefaces the organized outline)

    I. First main idea
       A. First subordinate idea
          1. Reason, example, or illustration
          2. Reason, example, or illustration
             a. Detail supporting reason #2
             b. Detail supporting reason #2
             c. Detail supporting reason #2
       B. Second subordinate idea
    II. Second main idea

Notice that each level of the paper must have more than one entry; for every A there must be at least a B (and, if required, a C, D, etc.), for every 1 there must be a 2. This arrangement forces you to compare ideas, looking carefully at each one to determine its place among the others. The insistence on assigning relative values to your ideas is what makes your outline an effective organizing tool.

THE PATTERNS OF CRIMINAL JUSTICE PAPERS

The structure of any particular type of criminal justice paper is governed by a formal pattern. When rigid external controls are placed on their writing, some writers tend to feel stifled, their creativity impeded by this kind of "paint-by-numbers" approach to structure. It is vital to the success of your paper that you never allow yourself to be overwhelmed by the pattern rules for any type of paper. Remember that such controls exist not to limit your creativity but to make the paper immediately and easily useful to its intended audience. It is as necessary to write clearly and confidently in a case study or a policy analysis paper as in a term paper for English literature, a résumé, a short story, or a job application letter.

Drafting

THE ROUGH DRAFT

The planning stage of the writing process is followed by the writing of the first draft. Using your thesis and outline as direction markers, you must now weave your amalgam of ideas, researched data, and persuasion strategies into logically ordered sentences and paragraphs. Though adequate prewriting may facilitate the drafting, it still will not be easy. Writers establish their own individual methods of encouraging themselves to forge ahead with the draft, but here are some tips to bear in mind.

1. Remember that this is a *rough draft,* not the final paper. At this stage, it is not necessary that every word be the best possible word. Do not put that sort of pressure on yourself. You must not allow anything to slow you down now. Writing is not like sculpting in stone, where every chip is permanent; you can always go back to your draft later and add, delete, reword, rearrange. *No matter how much effort you have put into planning, you cannot be sure how much of this first draft you will eventually keep.* It may take several drafts to get one that you find satisfactory.

2. Give yourself sufficient time to write. Don't delay the first draft by telling yourself there is still more research to do. You cannot uncover all the material there is to know on a particular subject, so don't fool yourself into trying. Remember that writing is a process of discovery. You may have to begin writing before you can see exactly what sort of final research you need to do. Keep in mind that there are other tasks waiting for you after the first draft is finished, so allow for them as you determine your writing schedule. It is also very important to give yourself time to write because the more time that passes after you have written a draft, the better your ability to view it with greater objectivity. It is very difficult to evaluate your writing accurately soon after you complete it. You need to cool down, to recover from the effort of putting all those words together. The "colder" you get on your writing, the better able you are to read it as if it were written by someone else and thus acknowledge the changes you will need to make to strengthen the paper.

3. Stay sharp. Keep in mind the plan you created for yourself as you narrowed your topic, composed a thesis sentence, and outlined the material. But if you begin to feel a strong need to change the plan a bit, do not be afraid to do so. Be ready for surprises dealt you by your own growing understanding of the topic. Your goal is to record your best thinking on the subject as accurately as possible.

LANGUAGE CHOICES

To be convincing, your writing needs to be authoritative. That is, you have to sound as if you have confidence in your ability to convey your ideas in words. Sentences that sound stilted or that suffer from weak phrasing or the use of clichés are not going to win supporters for the positions that you express in your paper. So a major question becomes: How can I sound confident? Here are some points to consider as you work to convey to your reader that necessary sense of authority.

LEVEL OF FORMALITY.   Tone is one of the primary methods by which you signal to the readers who you are and what your attitude is toward them and toward your topic. Your major decision is which level of language formality is most appropriate to your audience. The informal tone you would use in a letter to a friend might well be out of place in a paper on police corruption written for your criminal justice professor. Remember that tone is only part of the overall decision that you make about how to present your information. Formality is, to some extent, a function of individual word choices and phrasing. Is it appropriate to use contractions like *isn't* or *they'll?* Would the strategic use of a sentence fragment for effect be out of place? The use of informal language, the personal *I,* and the second person *you* is traditionally forbidden—for better or worse—in certain kinds of writing. Often part of the challenge of writing a formal paper is simply how to give your prose bite while staying within the conventions.

JARGON.   One way to lose readers quickly is to overwhelm them with jargon—phrases that have a special, usually technical, meaning within your discipline but which are un-familiar to the average reader. The occasional use of jargon may add an effective touch of atmosphere, but anything more than that will severely dampen a reader's enthusiasm for the paper. Often the writer uses jargon in an effort to impress the reader by sound-ing lofty or knowledgeable. Unfortunately, all jargon usually does is cause confusion. In fact, the use of jargon indicates a writer's lack of connection to the audience.

Criminal justice writing is a haven for jargon. Perhaps writers of professional jour-nals and certain policy analysis papers believe their readers are all completely attuned to their terminology. It may be that these writers occasionally hope to obscure damag-ing information or potentially unpopular ideas in confusing language. In other cases, the problem could simply be unclear thinking by the writer. Whatever the reason, the fact is that criminal justice papers too often sound like prose made by machines to be read by machines.

Students may feel that, in order to be accepted as criminal justice professionals, their papers should conform to the practices of their published peers. *This is a mistake.* Remember that it is never better to write a cluttered or confusing sentence than a clear one, and that burying your ideas in jargon defeats the effort that you went through to form them.

CLICHÉS.   In the heat of composition, as you are looking for words to help form your ideas, it is sometimes easy to plug in a *cliché*—a phrase that has attained universal recognition by overuse. (Note: Clichés differ from jargon in that clichés are part of the general public's everyday language, while jargon is specific to the language of experts in a particular field.) Our vocabularies are brimming with clichés:

- It's *raining cats and dogs.*
- That issue is *dead as a doornail.*
- It's time for the governor to *face the music.*
- Angry voters *made a beeline* for the ballot box.

Revising Jargon

What words in the following sentence, from a published article in a journal, are examples of jargon? Can you rewrite the sentence to clarify its meaning?

The implementation of statute-mandated regulated inputs exceeds the conceptual-ization of the administrative technicians.

The problem with clichés is that they are virtually meaningless. Once colorful means of expression, they have lost their color through overuse, and they tend to bleed energy and color from the surrounding words. When revising, replace clichés with wording that more accurately conveys your point.

DESCRIPTIVE LANGUAGE

Language that appeals to the readers' senses will always engage their interest more fully than language that is abstract. This is especially important for writing in disciplines that tend to deal in abstracts, such as criminal justice. The typical criminal justice paper, with its discussions of abstract principles, demographics, or deterministic outcomes, is usually in danger of floating off into abstraction, with each paragraph drifting farther away from the felt life of the readers. Whenever appropriate, appeal to your readers' sense of sight, hearing, taste, touch, or smell.

BIAS-FREE AND GENDER-NEUTRAL WRITING

Language can be a very powerful method of either reinforcing or destroying cultural stereotypes. By treating the sexes in subtly different ways in your language, you may unknowingly be committing an act of discrimination. A common example is the use of the pronoun *he* to refer to a person whose gender has not been identified. But there are many other writing situations in which sexist and/or ethnic bias may appear. In order to avoid gender bias, the *American Sociological Association's Style Guide* (1997) recommends replacing words like *man, men,* or *mankind* with *person, people,* or *humankind.* When both sexes must be referred to in a sentence, use *he or she, her or him,* or *his or hers* instead of *he/she, him/her,* or *his/hers.*

Some writers, faced with this dilemma, alternate the use of male and female personal pronouns; others use the plural to avoid the need to use a pronoun of either gender:

Sexist: A lawyer should always treat his client with respect.

Nonsexist: A lawyer should always treat his or her client with respect.

Nonsexist: Lawyers should always treat their clients with respect.

Sexist: Man is a political animal.

Nonsexist: People are political animals.

Using Descriptive Language

Which of the following two sentences is more effective?

1. The housing project had deteriorated since the last inspection.
2. Since the last inspection, deterioration of the housing project had become evident in the stench rising from the plumbing, grime on the walls and floors, and the sound of rats scurrying in the hallways.

Remember that language is more than the mere vehicle of your thought. Your words shape perceptions for your readers. How *well* you say something will profoundly affect your readers' response to *what* you say. Sexist language denies to a large number of your readers the basic right to fair and equal treatment. Be aware of this common form of discrimination.

Revising

Revising is one of the most important steps in assuring the success of your essay. While unpracticed writers often think of revision as little more than making sure all the *i*'s are dotted and *t*'s are crossed, it is much more than that. Revising is *re-seeing* the essay, looking at it from other perspectives, trying always to align your view with the one that will be held by your audience. Research indicates that we are actually revising all the time, in every phase of the writing process as we reread phrases, rethink the placement of an item in an outline, or test a new topic sentence for a paragraph. Subjecting your entire hard-fought draft to cold, objective scrutiny is one of the toughest activities to master, but it is absolutely necessary. You must make sure that you have said everything that needs to be said clearly and in logical order. One confusing passage, and the reader's attention is deflected from where you want it to be. Suddenly the reader has to become a detective, trying to figure out why you wrote what you did and what you meant by it. You do not want to throw such obstacles in the path of understanding. Here are some tips to help you with revision.

1. *Give yourself adequate time for revision.* As discussed above, you need time to become "cold" on your paper in order to analyze it objectively. After you have written your draft, spend some time away from it. When you return, try to think of it as someone else's paper.
2. *Read the paper carefully.* This is tougher than it sounds. One good strategy is to read it aloud or to have a friend read it aloud while you listen. (Note, however, that friends are usually not the best critics. They are rarely trained in revision techniques and are often unwilling to risk disappointing you by giving your paper a really thorough examination.)
3. *Have a list of specific items to check.* It is important to revise in an orderly fashion, in stages, looking first at large concerns, such as the overall structure, and then rereading for problems with smaller elements such as paragraph organization or sentence structure.
4. *Check for unity.* Unity is the clear and logical relation of all parts of the essay to its thesis. Make sure that every paragraph relates well to the whole of the paper and is in the right place.
5. *Check for coherence.* Make sure there are no gaps between the different parts of the argument. Look to see that you have adequate *transition* everywhere it is needed. Transitional elements are markers indicating places where the paper's focus or attitude changes. Transitional elements can be one word long—*however, although, unfortunately, luckily*—or as long as a sentence or a paragraph: *In order to appreciate fully the importance of democracy as a shaping presence in post–cold war Polish politics, it is necessary to examine briefly the Poles' last historical attempt to implement democratic government.*

Transitional elements rarely introduce new material. Instead, they are direction pointers, either indicating a shift to new subject matter or signaling how the writer wishes certain material to be interpreted by the reader. Because you, the writer, already know where and why your paper changes direction and how you want particular passages to be received, it can be very difficult for you to catch those places in your paper where transition is needed.

6. *Avoid unnecessary repetition.* Two types of repetition can annoy a reader: repetition of content and of wording.

Repetition of content occurs when you return to a subject that you have already discussed. Ideally, you should consider a topic *once,* memorably, and then move on to your next topic. Organizing a paper is a difficult task, however, that usually occurs through a process of enlightenment in terms of purposes and strategies, and repetition of content can happen even if you have used prewriting strategies. What is worse, it can be difficult for you to be aware of the repetition in your own writing. As you write and revise, remember that any unnecessary repetition of content in your final draft is potentially annoying to your readers, who are working to make sense of the argument they are reading and do not want to be distracted by a passage repeating material they have already encountered. You must train yourself, through practice, to look for material that you have repeated unnecessarily.

Repetition of wording occurs when you overuse certain phrases or words. This can make your prose sound choppy and uninspired, as the following examples demonstrate:

- The subcommittee's report on prison reform will surprise a number of people. A number of people will want copies of the report.
- The chairman said at a press conference that he is happy with the report. He will circulate it to the local news agencies in the morning. He will also make sure that the city council has copies.
- I became upset when I heard how the committee had voted. I called the chairman and expressed my reservations about the committee's decision. I told him I felt that he had let the teachers and students of the state down. I also issued a press statement.

The last passage illustrates a condition known by composition teachers as the *I-syndrome.* Can you hear how such duplicated phrasing can hurt a paper? Your language should sound fresh and energetic. Make sure, before you submit your final draft, to read through your paper carefully, looking for such repetition.

Not all repetition is bad. You may wish to repeat a phrase for rhetorical effect or special emphasis: *I came. I saw. I conquered.* Just make sure that any repetition in your paper is intentional, placed there to produce a specific effect.

Editing

Editing is sometimes confused with the more involved process of revising. But editing happens later, after you have wrestled through your first draft—and maybe your second and third—and arrived at the final draft. Even though your draft now contains all the information you want to impart and has arranged the information to your satisfaction, there are still many factors to check, such as sentence structure, spelling, and punctuation. It is at this point that an unpracticed writer might be less than vigilant.

After all, most of the work on the paper is finished; the big jobs of discovering material and organizing and drafting it have been completed. But watch out! Editing is as important as any other job in the writing process. Any error that you allow in the final draft will count against you in the mind of the reader. It may not seem fair, but a minor error—a misspelling or the confusing placement of a comma—will make a much greater impression on your reader than perhaps it should. Remember that everything about your paper is your responsibility, including getting even the supposedly little jobs right. Careless editing undermines the effectiveness of your paper. It would be a shame if all the hard work you put into prewriting, drafting, and revising were to be damaged because you carelessly allowed a comma splice!

Most of the tips given above for revising hold for editing as well. It is best to edit in stages, looking for only one or two kinds of errors each time you reread the paper. Focus especially on errors that you remember committing in the past. If, for instance, you know you have a tendency to misplace commas, go through your paper looking at each comma carefully. If you have a weakness for writing unintentional sentence fragments, read each sentence aloud to make sure that it is indeed a complete sentence. Have you accidentally shifted verb tenses anywhere, moving from past to present tense for no reason? Do all the subjects in your sentences agree in number with their verbs? Now is the time to find out.

Watch out for *miscues*—problems with a sentence that the writer simply does not see. Remember that your search for errors is hampered in two ways:

1. As the writer, you hope *not* to find any errors with your writing. This desire not to find mistakes can cause you to miss sighting them when they occur.
2. Since you know your material so well, it is easy as you read to supply a missing word or piece of punctuation *unconsciously,* as if it is present.

How difficult is it to see that something is missing in the following sentence:

```
Unfortunately, legislators often have too little regard
their constituents.
```

We can even guess that the missing word is probably *for,* which should be inserted after *regard.* It is quite possible, however, that the writer of the sentence would automatically supply the missing *for,* as if it were on the page. This is a miscue, which can be hard for writers to spot because they are so close to their material.

One tactic for catching mistakes in sentence structure is to read the sentences aloud, starting with the last one in the paper and then moving to the next-to-last, then the previous sentence, thus going backward through the paper (reading each sentence in the normal, left-to-right manner, of course) until you reach the first sentence of the introduction. This backward progression strips each sentence of its rhetorical context and helps you to focus on its internal structure.

Editing is the stage where you finally answer those minor questions that you put off earlier when you were wrestling with wording and organization. Any ambiguities regarding the use of abbreviations, italics, numerals, capital letters, titles (when do you capitalize *president,* for example?), hyphens, dashes (usually created on a typewriter or computer by striking the hyphen key twice), apostrophes, and quotation marks have to be cleared up now. You must also check to see that you have used the required formats for footnotes, endnotes, margins, and page numbers.

Guessing is not allowed. Sometimes unpracticed writers who realize that they don't quite understand a particular rule of grammar, punctuation, or format do nothing to fill that knowledge gap. Instead, they rely on guesswork and their own logic—which is not always up to the task of dealing with so contrary a language as English—to get them through problems that they could solve if only they referred to a writing manual. Remember that it does not matter to the reader why or how an error shows up in your writing. It only matters that you have dropped your guard. You must not allow a careless error to undo the good work that you have done.

Proofreading

Before you hand in your final version of the paper, it is vital that you check it over one more time to make sure there are no errors of any sort. This job is called *proofreading* or *proofing.* In essence, you are looking for many of the same things you checked for during editing, but now you are doing it on the last draft, which is about to be submitted to your audience. Proofreading is as important as editing; you may have missed an error that you still have time to find, or an error may have been introduced when the draft was recopied or typed for the last time. Like every other stage of the writing process, proofreading is your responsibility.

At this point, you must check for typing mistakes: transposed or deleted letters, words, phrases, or punctuation. If you have had the paper professionally typed, you still must check it carefully. Do not rely solely on the typist's proofreading. If you are creating your paper on a computer or a word processor, it is possible for you unintentionally to insert a command that alters your document drastically by slicing out a word, line, or sentence at the touch of a key. Make sure such accidental deletions have not occurred.

Above all else, remember that your paper represents you. It is a product of your best thinking, your most energetic and imaginative response to a writing challenge. If you have maintained your enthusiasm for the project and worked through the different stages of the writing process honestly and carefully, you should produce a paper you can be proud of and one that will serve its readers well.

Writing Competently

GUIDELINES FOR THE COMPETENT WRITER

Good writing places your thoughts in your readers' minds in exactly the way you want them to be there. It tells your readers just what you want them to know without telling them anything you do not wish to say. That may sound odd, but the fact is that writers have to be careful not to let unwanted messages slip into their writing. Look, for example, at the passage below, taken from a paper analyzing the impact of a worker-retraining program in the writer's state. Hidden within the prose is a message that jeopardizes the paper's success. Can you detect the message?

> *Recent articles written on the subject of dislocated workers have had little to say about the particular problems dealt with in this paper. Since few of these articles focus on the problem at the local level.*

Chances are, when you reached the end of the second "sentence," you felt that something was missing and perceived a gap in logic or coherence, so you went back through both sentences to find the place where things had gone wrong. The second sentence is actually not a sentence at all. It does have certain features of a sentence—a subject *(few)*, and a verb *(focus)*—but its first word *(Since)* subordinates the entire clause that follows, taking away its ability to stand on its own as a complete idea. The second "sentence," which is properly called a *subordinate clause,* merely fills in some information about the first sentence, telling us why recent articles about dislocated workers fail to deal with problems discussed in the present paper.

The sort of error represented by the second "sentence" is commonly called a *sentence fragment,* and it conveys to the reader a message that no writer wants to send: that the writer either is careless or—worse—has not mastered the language. Language errors such as fragments, misplaced commas, or shifts in verb tense send out warnings in the readers' minds. As a result the readers lose a little of their concentration on the issue being discussed. They become distracted and begin to wonder about the language competency of the writer. The writing loses effectiveness.

NOTE: Whatever goal you set for your paper, whether you want it to persuade, describe, analyze, or speculate, you must also set another goal: to *display language competence.* If your paper does not meet this goal, it will not completely achieve its other aims. Language errors spread doubt like a virus; they jeopardize all the hard work you have done on your paper.

Credibility in the job market depends upon language competence. Anyone who doubts this should remember the beating that Vice President Dan Quayle took in the press for misspelling the word *potato* at a spelling bee in 1992. His error caused a storm of humiliating publicity for the hapless Quayle, adding to an impression of his general incompetence.

Correctness is Relative. Although they may seem minor, the sort of language errors we are discussing—often called *surface errors*—can be extremely damaging in certain kinds of writing. Surface errors come in a variety of types, including misspellings, punctuation problems, grammar errors, and the inconsistent use of abbreviations, capitalization, or numerals. These errors are an affront to your reader's notion of correctness, and therein lies one of their biggest problems. Different audiences tolerate different levels of correctness. You know that you can get away with surface errors in, say, a letter to a friend, who will not judge you harshly for them, while those same errors in a job application letter might eliminate you from consideration for the position. Correctness depends to an extent upon context.

Another problem is that the rules governing correctness shift over time. What would have been an error to your grandmother's generation—the splitting of an infinitive, for example, or the ending of a sentence with a preposition—is taken in stride today by most readers. So how do you write correctly when the rules shift from person to person and over time? Here are some tips.

Consider Your Audience

One of the great risks of writing is that even the simplest choices regarding wording or punctuation can sometimes prejudice your audience against you in ways that may seem unfair. For example, look again at the old grammar rule forbidding the splitting of infinitives. After decades of counseling students to *never* split an infinitive (something this sentence has just done), composition experts now concede that a split infinitive is not a grammar crime. But suppose you have written a position paper trying to convince your city council of the need to hire security personnel for the library, and half of the council members—the people you wish to convince—remember their eighth-grade grammar teacher's outdated warning about splitting infinitives. How will they respond when you tell them, in your introduction, that librarians are compelled "to always accompany" visitors to the rare book room because of the threat of vandalism? How much of their attention have you suddenly lost because of their automatic recollection of what is now a nonrule? It is possible, in other words, to write correctly and still offend your readers' notions of language competence.

Make sure that you tailor the surface features and the degree of formality of your writing to the level of competency that your readers require. When in doubt, take a conservative approach. Your audience might be just as distracted by contractions as by a split infinitive.

Aim for Consistency

When dealing with a language question for which there are different answers—such as whether or not to place a comma after the second item in a series of three ("The mayor's speech addressed taxes, housing for the poor, and the job situation")—always use the same strategy. If, for example, you avoid splitting one infinitive, avoid splitting all infinitives in your paper.

Have Confidence in What You Already Know About Writing!

It is easy for inexperienced writers to allow their occasional mistakes to shake their confidence in their writing ability. The fact is, however, most of what we know about writing is correct. We are all capable, for example, of writing grammatically sound phrases, even if we cannot list the rules by which we achieve coherence. Most writers who worry about their chronic errors have fewer than they think. Becoming distressed about errors makes writing more difficult.

Eliminate Chronic Errors

But if just thinking about our errors has a negative effect on our writing, how do we learn to write more correctly? Perhaps the best answer is simply to write as often as possible. Give yourself practice in putting your thoughts into written shape—and get lots of practice in revising and proofing your work. And as you write and revise, be honest with yourself—and patient. Chronic errors are like bad habits; getting rid of them takes time.

Grammar

As various composition theorists have pointed out, the word *grammar* has several definitions. One meaning is "the formal patterns in which words must be arranged in order to convey meaning." We learn these patterns very early in life and use them spontaneously without thinking about them. Our understanding of grammatical patterns is extremely sophisticated, despite the fact that few of us can actually cite the rules by which the patterns work. Hartwell (1985:111) tested grammar learning by asking native English speakers of different ages and levels of education, including high school teachers, to arrange these words in natural order:

French   the   young   girls   four

Everyone could produce the natural order for this phrase: "the four young French girls." Yet none of Hartwell's respondents said they knew the rule that governs the order of the words.

You probably know of one or two problem areas in your writing that you could have eliminated but have not done so. Instead, you have "fudged" your writing at the critical points, relying upon half-remembered formulas from past English classes or trying to come up with logical solutions to your writing problems. *(Warning:* The English language does not always work in a way that seems logical.) You may have simply decided that comma rules are unlearnable or that you will never understand the difference between the verbs *lay* and *lie.* And so you guess, and come up with the wrong answer a good part of the time. What a shame, when just a little extra work would give you mastery over those few gaps in your understanding and boost your confidence as well.

Instead of continuing with this sort of guesswork and living with the gaps in your knowledge, why not face the problem areas now and learn the rules that have heretofore escaped you? What follows is a discussion of those surface features of writing where errors most commonly occur. You will probably be familiar with most if not all of the rules discussed, but there may well be a few you have not yet mastered. Now is the time to do so.

PUNCTUATION, GRAMMAR, AND SPELLING

Apostrophes

An apostrophe is used to show possession; when you wish to say that something belongs to someone or to something, you add either an apostrophe and an *s* or an apostrophe alone to the word that represents the owner.

When the owner is *singular* (a single person or thing), the apostrophe precedes an added *s:*

- According to Mr. Pederson's secretary, the board meeting has been canceled.
- The school's management team reduced crime problems last year.
- Somebody's briefcase was left in the classroom.

The same rule applies if the word showing possession is a plural that does not end in *s:*

- The women's club provided screening services for at-risk youth and their families.
- Professor Logan has proven himself a tireless worker for children's rights.

When the word expressing ownership is a *plural* ending in *s,* the apostrophe follows the *s:*

- The new procedure was discussed at the youth workers' conference.

There are two ways to form the possessive for two or more nouns:

1. To show joint possession (both nouns owning the same thing or things), the last noun in the series is possessive: Billy and Richard's first draft was completed yesterday.
2. To indicate that each noun owns an item or items individually, each noun must show possession: Professor Wynn's and Professor Camacho's speeches took different approaches to the same problem.

The apostrophe is important, an obvious statement when you consider the difference in meaning between the following two sentences:

1. Be sure to pick up the psychiatrist*'s* things on your way to the airport.
2. Be sure to pick up the psychiatrist*s'* things on your way to the airport.

In the first of these sentences, you have only one psychiatrist to worry about, while in the second, you have at least two!

Capitalization

WHEN TO CAPITALIZE

Here is a brief summary of some hard-to-remember capitalization rules.

1. You may, if you choose, capitalize the first letter of the first word in a sentence that follows a colon, but you do not have to do so. Make sure, however, that you use one pattern consistently throughout your paper:

   - Our instructions are explicit: Do not allow anyone into the conference without an identification badge.
   - Our instructions are explicit: do not allow anyone into the conference without an identification badge.

2. Capitalize *proper nouns* (nouns naming specific people, places, or things) and *proper adjectives* (adjectives made from proper nouns). A common noun following the proper adjective is usually not capitalized, nor is a common article preceding the proper adjective (such as *a*, *an*, or *the*):

Proper Nouns	Proper Adjectives
England	English sociologists
Iraq	the Iraqi educator
Shakespeare	a Shakespearean tragedy

Proper nouns include:

- *Names of monuments and buildings:* the Washington Monument, the Empire State Building, the Library of Congress
- *Historical events, eras, and certain terms concerning calendar dates:* the Civil War, the Dark Ages, Monday, December, Columbus Day
- *Parts of the country:* North, Southwest, Eastern Seaboard, the West Coast, New England

NOTE: When words like *north, south, east, west, northwest* are used to designate direction rather than geographical region, they are not capitalized: We drove *east* to Boston and then made a tour of the *East Coast.*

- *Words referring to race, religion, or nationality:* Islam, Muslim, Caucasian, Asian, African American, Slavic, Arab, Jewish, Hebrew, Buddhism, Buddhists, Southern Baptists, the Bible, the Koran, American

- *Names of languages:* English, Chinese, Latin, Sanskrit
- *Titles of corporations, institutions, businesses, universities,* and *organizations:* Dow Chemical, General Motors, the National Endowment for the Humanities, University of Tennessee, Colby College, Kiwanis Club, American Association of Retired Persons, the Oklahoma State Senate

NOTE: Some words once considered proper nouns or adjectives have, over time, become common, such as french fries, pasteurized milk, arabic numerals, and italics.

3. Titles of individuals may be capitalized if they precede a proper name; otherwise, titles are usually not capitalized.

- The committee honored Dean Furmanski.
- The committee honored the deans from the other colleges.
- We phoned Doctor MacKay, who arrived shortly afterward.
- We phoned the doctor, who arrived shortly afterward.
- A story on Queen Elizabeth's health appeared in yesterday's paper.
- A story on the queen's health appeared in yesterday's paper.
- Pope John Paul's visit to Colorado was a public relations success.
- The pope's visit to Colorado was a public relations success.

WHEN NOT TO CAPITALIZE

In general, you do not capitalize nouns when your reference is nonspecific. For example, you would not capitalize the phrase *the senator,* but you would capitalize *Senator Smith.* The second reference is as much a title as it is a mere term of identification, while the first reference is a mere identifier. Likewise, there is a difference in degree of specificity between the phrase *the state treasury* and *the Texas State Treasury.*

NOTE: The meaning of a term may change somewhat depending on capitalization. What, for example, might be the difference between a *Democrat* and a *democrat?* When capitalized, the word refers to a member of a specific political party; when not capitalized, it refers to someone who believes in a democratic form of government.

Capitalization depends to some extent on the context of your writing. For example, if you are writing a policy analysis for a specific corporation, you may capitalize words and phrases referring to that corporation—such as *Board of Directors, Chairman of the Board,* and *the Institute*—that would not be capitalized in a paper written for a more general audience. Likewise, in some contexts it is not unusual to see titles of certain powerful officials capitalized even when not accompanying a proper noun: The President's visit to the Oklahoma City bombing site was considered a success.

Colons

We all know certain uses for the colon. A colon can, for example, separate the parts of a statement of time (4:25 A.M.), separate chapter and verse in a biblical quotation (John 3:16), and close the salutation of a business letter (Dear Senator Keaton:). But the colon has other uses that can add an extra degree of flexibility to sentence structure.

The colon can introduce into a sentence certain kinds of material, such as a list, a quotation, or a restatement or description of material mentioned earlier:

- *List:* The committee's research proposal promised to do three things: (1) establish the extent of the problem, (2) examine several possible solutions, and (3) estimate the cost of each solution.
- *Quotation:* In his speech, the mayor challenged us with these words: "How will your council's work make a difference in the life of our city?"
- *Restatement or description:* Ahead of us, according to the senator's chief of staff, lay the biggest job of all: convincing our constituents of the plan's benefits.

Commas

The comma is perhaps the most troublesome of all marks of punctuation, no doubt because its use is governed by so many variables, such as sentence length, rhetorical emphasis, and changing notions of style. The most common problems are outlined below.

THE COMMA SPLICE

A comma splice is the joining of two complete sentences by only a comma:

- An impeachment is merely an indictment of a government official, actual removal usually requires a vote by a legislative body.
- An unemployed worker who has been effectively retrained is no longer an economic problem for the community, he has become an asset.
- It might be possible for the city to assess fees on the sale of real estate, however, such a move would be criticized by the community of real estate developers.

In each of these passages, two complete sentences (also called *independent clauses)* have been spliced together by a comma, which is an inadequate break between the two sentences.

One foolproof way to check your paper for comma splices is to read carefully the structures on both sides of each comma. If you find a complete sentence on each side, and if the sentence following the comma does not begin with a coordinating conjunction *(and, but, for, nor, or, so, yet),* then you have found a comma splice.

Simply reading the draft through to try to "hear" the comma splices may not work, since the rhetorical features of your prose—its "movement"—may make it hard to detect this kind of sentence completeness error. There are five commonly used ways to correct comma splices.

1. Place a period between the two independent clauses:

    Incorrect: Physicians receive many benefits from their affiliation with clients, there are liabilities as well.

    Correct: Physicians receive many benefits from their affiliation with clients. There are liabilities as well.

2. Place a comma and a coordinating conjunction *(and, but, for, or, nor, so, yet)* between the sentences:

> *Incorrect:*   The chairperson's speech described the major differences of opinion over the department situation, it also suggested a possible course of action.
>
> *Correct:*   The chairperson's speech described the major differences of opinion over the departmental situation, and it also suggested a possible course of action.

3. Place a semicolon between the independent clauses:

> *Incorrect:*   Some people believe that the federal government should play a large role in establishing a housing policy for the homeless, many others disagree.
>
> *Correct:*   Some people believe that the federal government should play a large role in establishing a housing policy for the homeless; many others disagree.

4. Rewrite the two clauses of the comma splice as one independent clause:

> *Incorrect:*   Television programs play a substantial part in the development of delinquent attitudes, however, they were not found to be the deciding factor in determining the behavior of juvenile delinquents.
>
> *Correct:*   Television programs were found to play a substantial but not a decisive role in determining the delinquent behavior of juveniles.

5. Change one of the two independent clauses into a dependent clause by beginning it with a *subordinating word (although, after, as, because, before, if, though, unless, when, which, where),* which prevents the clause from being able to stand on its own as a complete sentence.

> *Incorrect:*   The student meeting was held last Tuesday, there was a poor turnout.
>
> *Correct:*   When the student meeting was held last Tuesday, there was a poor turnout.

COMMAS IN A COMPOUND SENTENCE

A compound sentence is comprised of two or more independent clauses—two complete sentences. When these two clauses are joined by a coordinating conjunction, the conjunction should be preceded by a comma to signal the reader that another independent clause follows. (This is the second method for fixing a comma splice described above.) When the comma is missing, the reader is not expecting to find the second half of a compound sentence and may be distracted from the text.

As the following examples indicate, the missing comma is especially a problem in longer sentences or in sentences in which other coordinating conjunctions appear.

Notice how the comma sorts out the two main parts of the compound sentence, eliminating confusion:

- *Without the comma:* The senator promised to visit the hospital and investigate the problem and then he called the press conference to a close.
- *With the comma:* The senator promised to visit the hospital and investigate the problem, and then he called the press conference to a close.
- *Without the comma:* The water board can neither make policy nor enforce it nor can its members serve on auxiliary water committees.
- *With the comma:* The water board can neither make policy nor enforce it, nor can its members serve on auxiliary water committees.

An exception to this rule arises in shorter sentences, where the comma may not be necessary to make the meaning clear:

- The mayor phoned and we thanked him for his support.

However, it is never wrong to place a comma before the conjunction between the independent clauses. If you are the least bit unsure of your audience's notions about what makes for "proper" grammar, it is a good idea to take the conservative approach and use the comma:

- The mayor phoned, and we thanked him for his support.

COMMAS WITH RESTRICTIVE AND NONRESTRICTIVE ELEMENTS

A *nonrestrictive element* is part of a sentence—a word, phrase, or clause—that adds information about another element in the sentence without restricting or limiting its meaning. While this information may be useful, the nonrestrictive element is not needed for the sentence to make sense. To signal its inessential nature, the nonrestrictive element is set off from the rest of the sentence with commas.

The failure to use commas to indicate the nonrestrictive nature of a sentence element can cause confusion. See, for example, how the presence or absence of commas affects our understanding of the following sentence:

1. The judge was talking with the policeman, who won the outstanding service award last year.
2. The judge was talking with the policeman who won the outstanding service award last year.

Can you see that the comma changes the meaning of the sentence? In the first version, the comma makes the information that follows it incidental: *The judge was talking with the policeman, who happens to have won the service award last year.* In the second version, the information following the word *policeman* is important to the sense of the sentence; it tells us specifically *which* policeman—presumably there are more than one—the judge was addressing. Here the lack of a comma has transformed the material following the word *policeman* into a *restrictive element,* which means that it is necessary to our understanding of the sentence.

Be sure that in your paper you make a clear distinction between nonrestrictive and restrictive elements by setting off the nonrestrictive elements with commas.

COMMAS IN A SERIES

A series is any two or more items of a similar nature that appear consecutively in a sentence. The items may be individual words, phrases, or clauses. In a series of three or more items, the items are separated by commas:

- *The senator, the mayor,* and *the police chief* all attended the ceremony.
- Because of the new zoning regulations, *all trailer parks must be moved out of the neighborhood, all small businesses must apply for recertification and tax status,* and *the two local churches must repave their parking lots.*

The final comma in the series, the one before the *and,* is sometimes left out, especially in newspaper writing. This practice, however, can make for confusion, especially in longer complicated sentences, like the second example above. Here is the way that sentence would read without the final, or *serial,* comma:

- Because of the new zoning regulations, all trailer parks must be moved out of the neighborhood, all small businesses must apply for recertification and tax status and the two local churches must repave their parking lots.

Notice that without a comma the division between the second and third items in the series is not clear. This is the sort of ambiguous structure that can cause a reader to backtrack and lose concentration. You can avoid such confusion by always using that final comma. Remember, however, that if you do decide to include it, do so *consistently;* make sure it appears in every series in your paper.

Dangling Modifiers

A *modifier* is a word or group of words used to describe, or modify, another word in the sentence. A *dangling modifier* appears either at the beginning or end of a sentence and seems to be describing some word other than the one the writer obviously intended. The modifier therefore "dangles," disconnected from its correct meaning. It is often hard for the writer to spot a dangling modifier, but readers can—and will—find them, and the result can be disastrous for the sentence, as the following examples demonstrate:

Incorrect:	Flying low over Washington, the White House was seen.
Correct:	Flying low over Washington, we saw the White House.
Incorrect:	Worried at the cost of the program, sections of the bill were trimmed in committee.
Correct:	Worried at the cost of the program, the committee trimmed sections of the bill.
Incorrect:	To lobby for prison reform, a lot of effort went into the TV ads.
Correct:	The lobby group put a lot of effort into the TV ads advocating prison reform.
Incorrect:	Stunned, the TV broadcast the defeated senator's concession speech.
Correct:	The TV broadcast the stunned senator's concession speech.

Note that in the first two incorrect sentences above, the confusion is largely due to the use of *passive-voice verbs:* "the prison *was seen,*" "sections of the proposal *were trimmed.*" Often, though not always, a dangling modifier results from the fact that the actor in the sentence—*we* in the first sentence, *the committee* in the second—is either distanced from the modifier or obliterated by the passive-voice verb. It is a good idea to avoid passive voice unless you have a specific reason for using it.

One way to check for dangling modifiers is to examine all modifiers at the beginnings or endings of your sentences. Look especially for *to be* phrases *(to lobby)* or for words ending in *-ing* or *-ed* at the start of the modifier. Then check to see if the word being modified is always in plain sight and close enough to the phrase to be properly connected.

Parallelism

Series of two or more words, phrases, or clauses within a sentence should have the same grammatical structure, which is called *parallelism.* Parallel structures can add power and balance to your writing by creating a strong rhetorical rhythm. Here is a famous example of parallelism from the Preamble to the U.S. Constitution. (The capitalization follows that of the original eighteenth-century document; parallel structures have been italicized.)

> We the People of the United States, in Order *to form a more perfect Union, Establish Justice, insure Domestic Tranquillity, provide for the common defense, promote the general Welfare, and secure the Blessings of Liberty to ourselves and our Posterity,* do *ordain and establish* this Constitution for the United States of America.

There are actually two series in this sentence, the first composed of six phrases that each complete the infinitive phrase beginning with the word *to (to form, [to] Establish, [to] insure, [to] provide, [to] promote, [to] secure),* the second consisting of two verbs *(do ordain and [do] establish).* These parallel series appeal to our love of balance and pattern, and they give an authoritative tone to the sentence. The writer, we feel, has thought long and carefully about the matter at hand and has taken firm control of it.

Because we find a special satisfaction in balanced structures, we are more likely to remember ideas phrased in parallelisms than in less highly ordered language. For this reason, as well as for the sense of authority and control that they suggest, parallel structures are common in well-written speeches:

> We hold these truths to be self-evident, that all men are created equal, that they are endowed by their Creator with certain unalienable Rights, that among these are Life, Liberty, and the pursuit of Happiness.
>
> > Declaration of Independence, 1776

> But, in a larger sense, we can not dedicate—we can not consecrate—we can not hallow—this ground. The brave men, living and dead, who struggled here, have consecrated it, far above our poor power to add or detract. The world will little note, nor long remember what we say here, but it can never forget what they did here.
>
> > Abraham Lincoln, Gettysburg Address, 1863

Let us never negotiate out of fear. But never let us fear to negotiate. . . . . Ask not what your country can do for you; ask what you can do for your country.
John F. Kennedy, Inaugural Address, 1961

FAULTY PARALLELISM

If the parallelism of a passage is not carefully maintained, the writing can seem sloppy and out of balance. Scan your writing to make sure that all series and lists have parallel structure. The following examples show how to correct faulty parallelism:

Incorrect:	The mayor promises not only to reform the police department, but also *the giving of raises* to all city employees. [Connective structures such as *not only . . . but also,* and *both . . . and* introduce elements that should be parallel.]
Correct:	The mayor promises not only *to reform* the police department, but also *to give* raises to all city employees.
Incorrect:	The cost *of doing* nothing is greater than the cost *to renovate* the apartment block.
Correct:	The cost of *doing* nothing is greater than the cost *of renovating* the apartment block.
Incorrect:	Here are the items on the committee's agenda: (1) *to discuss* the new property tax, (2) *to revise* the wording of the city charter, (3) *a vote* on the city manager's request for an assistant.
Correct:	Here are the items on the committee's agenda: (1) *to discuss* the new property tax, (2) *to revise* the wording of the city charter, (3) *to vote* on the city manager's request for an assistant.

Fused (Run-on) Sentences

A *fused sentence* is one in which two or more independent clauses (passages that can stand as complete sentences) have been joined together without the aid of any suitable connecting word, phrase, or punctuation. The sentences have been run together. As you can see, there are several ways to correct a fused sentence:

Incorrect:	The council members were exhausted they had debated for two hours.
Correct:	The council members were exhausted. They had debated for two hours. [The linked independent clauses have been separated into two sentences.]
Correct:	The council members were exhausted; they had debated for two hours. [A semicolon marks the break between the two clauses.]
Correct:	The council members were exhausted, having debated for two hours. [The second independent clause has been rephrased as a dependent clause.]
Incorrect:	Our policy analysis impressed the committee it also convinced them to reconsider their action.

> *Correct*:   Our policy analysis impressed the committee and also convinced them to reconsider their action. [The second clause has been re-phrased as part of the first clause.]
>
> *Correct:*   Our policy analysis impressed the committee, and it also convinced them to reconsider their action. [The two clauses have been separated by a comma and a coordinating word.]

Although a fused sentence is easily noticeable to the reader, it can be maddeningly difficult for the writer to catch in proofreading. Unpracticed writers tend to read through the fused spots, sometimes supplying the break that is usually heard when sentences are spoken. To check for fused sentences, read the independent clauses in your paper carefully, making sure that there are adequate breaks among all of them.

Pronoun Errors

ITS VERSUS IT'S

Do not make the mistake of trying to form the possessive of it in the same way that you form the possessive of most nouns. The pronoun it shows possession by simply adding an s:

- The prosecuting attorney argued the case on *its* merits.

The word *it's* is a contraction, meaning *it is:*

- *It's* the most expensive program ever launched by the prison.

What makes the *its/it's* rule so confusing is that most nouns form the singular possessive by adding an apostrophe and an *s:*

- The *jury's* verdict startled the crowd.

When proofreading, any time you come to the word *it's,* substitute the phrase *it is* while you read. If the phrase makes sense, you have used the correct form. If you have used the word *it's:*

- The newspaper article was misleading in *it's* analysis of the election.

Then read it as *it is:*

- The newspaper article was misleading in *it is* analysis of the election.

If the phrase makes no sense, substitute *its* for *it's:*

- The newspaper article was misleading in *its* analysis of the election.

VAGUE PRONOUN REFERENCE

Pronouns are words that stand in place of nouns or other pronouns that have already been mentioned in your writing. The most common pronouns include *he, she, it, they, them, those, which,* and *who.* You must make sure that there is no confusion about the word to which each pronoun refers:

- The mayor said that *he* would support our bill if the city council would also back *it.*
- The piece of legislation *which* drew the most criticism was the bill concerning housing for the poor.

The word that is replaced by the pronoun is called its antecedent. To check the accuracy of your pronoun references, ask yourself this question: To what does the pronoun refer? Then answer the question carefully, making sure that there is not more than one possible antecedent. Consider the following example:

- Several special interest groups decided to defeat the new health care bill. *This* became the turning point of the government's reform campaign.

To what does the word *This* refer? The immediate answer seems to be the word bill at the end of the previous sentence. It is more likely the writer was referring to the attempt of the special interest groups to defeat the bill, but there is no word in the first sentence that refers specifically to this action. The reference is unclear. One way to clarify the reference is to change the beginning of the second sentence:

- Several special interest groups decided to defeat the new health care bill. *Their attack on the bill* became the turning point of the government's reform campaign.

Here is another example:

- When John F. Kennedy appointed his brother Robert to the position of U.S. Attorney General, *he* had little idea how widespread the corruption in the Teamsters Union was.

To whom does the word *he* refer? It is unclear whether the writer is referring to John or to Robert Kennedy. One way to clarify the reference is simply to repeat the antecedent instead of using a pronoun:

- When President John F. Kennedy appointed his brother Robert to the position of U. S. Attorney General, *Robert* had little idea how widespread the corruption in the Teamsters Union was.

PRONOUN AGREEMENT

Remember that a pronoun must agree with its antecedent in both gender and number, as the following examples demonstrate:

- Mayor Smith said that *he* appreciated our club's support in the election.
- One reporter asked the senator what *she* would do if the President offered *her* a cabinet post.
- Having listened to our case, the judge decided to rule on *it* within the week.
- Engineers working on the housing project said *they* were pleased with the renovation so far.

The following words, however can become troublesome antecedents. They may look like plural pronouns but are actually singular:

Anyone	Each	Either	Everybody	Everyone
Nobody	No one	Somebody	Someone	

A pronoun referring to one of these words in a sentence must be singular, too.

Incorrect: *Each* of the women in the support group brought *their* children.

Correct: *Each* of the women in the support group brought *her* children.

Incorrect: Has *everybody* received *their* ballot?

Correct: Has *everybody* received *his or her* ballot? [The two gender-specific pronouns are used to avoid sexist language.]

Correct: Have *all* the delegates received *their* ballots? [The singular antecedent has been changed to a plural one.]

SHIFT IN PERSON

It is important to avoid shifting among first person (I, we), second person (you), and third person (she, he, it, one, they) unnecessarily. Such shifts can cause confusion:

Incorrect: *Most people* [third person] who seek a job find that if *you* [second person] tell the truth during *your* interviews, *you* will gain the voters' respect.

Correct: *Most people* who seek a job find that if *they* tell the truth during *their* interviews, *they* will win the voters' respect.

Incorrect: *One* [third person singular] cannot tell whether *they* [third person plural] are cut out for public office until *they* decide to run.

Correct: *One* cannot tell whether *one* is cut out for public office until *one* decides to run.

Quotation Marks

It can be difficult to remember when to use quotation marks and where they go in relation to other marks of punctuation. When faced with these questions, inexperienced often try to rely on logic rather than on a rulebook, but the rules do not always seem to rely on logic. The only way to make sure of your use of quotation marks is to *memorize* the rules. Luckily, there are not many.

QUOTATION MARKS AND DIRECT QUOTATIONS

Use quotation marks to enclose direct quotations that are not longer than four typed lines:

It remains for history to pass judgment on what one scholar calls "the strangest verdict ever given in such a case" (Rollins 2001:45).

Longer quotes, called *block quotes,* are handled in different ways according to the style guide you are using. For example, the *American Sociological Association's Style Guide,* which is often used by criminal justice professionals, requires that block quotes appear, in smaller type, in an indented block without quotation marks:

Sowell's position is summarized by Scott (1997):

> The constrained vision sees people as fundamentally limited in terms of their abilities to live peaceful, cooperative, public-spirited lives. People are morally limited, and therefore, although they may do a good deed, they are not to be trusted to act as they ought to toward one another. If we hear of a disaster in another part of the world, we may take a moment to feel sorry for the victims, but then we proceed with our lives as if nothing had happened. (P. 46)

In this example, the author's name and the date of publication appear within the paper's text, while the page number of the quote is given, in parentheses, following the quote. (Note that the *p* representing *page number* is capitalized when it is the first item in the parentheses.) You may, if you wish, include author's name and date within the parentheses instead of in your text.

The block quote format in the *Publication Manual of the American Psychological Association,* another style guide often used by criminal justice professionals, requires quotations of forty words or longer to be indented and presented without quotation marks just as the *ASA Style Guide* does, but does not require a change of type size for the quote. Also, the APA manual does not capitalize the *p* representing *page number* if it appears at the beginning of the parenthetical reference following the quote: (p. 46)

NOTE: Bibliographical formatting styles for both the *ASA Style Guide* and the *Publication Manual of the APA* are given in Chapter 4. Whichever bibliographical format you use, *be consistent.*

Use single quotation marks to set off quotations within quotations:

- "One of the defendants objected," Officer Smith explained, "saying that he refused 'to acknowledge any wrongdoing in the case whatsoever.' "

When the interior quote occurs at the end of the sentence, both single and double quotation marks are placed outside the period.

Use quotation marks to set off titles of the following:

- a short poem (one not printed as a separate volume)
- a short story
- an article or essay
- a song title
- an episode of a television or radio show

Use quotation marks to set off words or phrases used in special ways:

- *To convey irony:* The "liberal" administration has done nothing but cater to big business.
- *To set off a technical term:* To have "charisma," Weber would argue, is to possess special powers. Many believe that John F. Kennedy had great charisma.

NOTE: Once the term is defined, it is not placed in quotation marks again.

QUOTATION MARKS IN RELATION TO OTHER PUNCTUATION

Always place commas and periods inside closing quotation marks:

- "My fellow Americans," said the President, "there are tough times ahead of us."

  Place colons and semicolons *outside* closing quotation marks:

- In his speech on voting, the sociologist warned against "an encroaching indolence"; he was referring to the middle class.
- There are several victims of the government's campaign to "Turn Back the Clock": the homeless, the elderly, and the mentally impaired.

Use the context to determine whether to place question marks, exclamation points, and dashes inside or outside closing quotation marks. If the punctuation is part of the quotation, place it *inside* the quotation mark:

- "When will the tenure committee make up its mind?" asked the dean.
- The demonstrators shouted, "Free the hostages!" and "No more slavery!"

If the punctuation is not part of the quotation, place it *outside* the quotation mark:

- Which president said, "We have nothing to fear but fear itself"? [Although the quote is a complete sentence, you do not place a period after it. There can only be one piece of "terminal" punctuation that ends a sentence.]

Semicolons

The semicolon is another little used punctuation mark you should learn to incorporate into your writing strategy because of its many potential applications. For example, a semicolon can be used to correct a comma splice:

Incorrect:	The union representatives left the meeting in good spirits, their demands were met.
Correct:	The union representatives left the meeting in good spirits; their demands were met.
Incorrect:	Several guests at the fund-raiser had lost their invitations, however, we were able to seat them anyway.
Correct:	Several guests at the fund-raiser had lost their invitations; however, we were able to seat them anyway. [Conjunctive adverbs like *however, therefore,* and *thus* are not coordinating words (such as and, but, or, for, so, yet) and cannot be used with a comma to link independent clauses. If the second independent clause begins with *however,* it must be preceded by either a period or a semicolon.]

As you can see from the second example above, connecting the two independent clauses with a semicolon instead of a period strengthens the relationship between the clauses.

Semicolons can also separate items in a series when the series itself contain commas:

- The newspaper account of the rally stressed the march, which drew the biggest crowd; the mayor's speech, which drew tremendous applause; and the party afterwards in the park.

Avoid misusing semicolons. For example, use a comma, not a semicolon, to separate an independent clause from a dependent clause:

Incorrect:	Students from the college volunteered to answer phones during the pledge drive; which was set up to generate money for the new arts center.
Correct:	Students from the college volunteered to answer phones during the pledge drive, which was set up to generate money for the new arts center.

Do not overuse semicolons. Although they are useful, too many semicolons in your writing can distract your readers' attention. Avoid monotony by using semicolons sparingly.

Sentence Fragments

A *fragment* is a part of a sentence that is punctuated and capitalized as if it were an entire sentence. It is an especially disruptive kind of error because it obscures the connections that the words of a sentence must make in order to complete the reader's understanding.

Students sometimes write fragments because they are concerned that a particular sentence is growing too long and needs to be shortened. Remember that cutting the length of a sentence merely by adding a period somewhere along its length often creates a fragment. When checking your writing for fragments, it is essential that you read each sentence carefully to determine whether it has (1) a complete subject and a verb, and (2) a subordinating word before the subject and verb, which makes the construction a subordinate clause rather than a complete sentence.

TYPES OF SENTENCE FRAGMENTS

Some fragments lack a verb:

Incorrect:	The chairperson of our committee, having received a letter from the mayor. [Note that the word *having,* which can be used as a verb, is here being used as a gerund introducing a participial phrase. *Watch out* for words that look like verbs but are being used in another way.]
Correct:	The chairperson of our committee received a letter from the mayor.

Some fragments lack a subject:

Incorrect:	Our study shows that there is broad support for improvement in the health care system. And in the unemployment system.

Correct: Our study shows that there is broad support for improvement in the health care system and in the unemployment system.

Some fragments are subordinate clauses:

Incorrect: After the latest edition of the newspaper came out. [This clause has the two major components of a complete sentence: a subject *(edition)* and a verb *(came)*. Indeed, if the first word *(After)* were deleted, the clause would be a complete sentence. But that first word is a *subordinating word,* which acts to prevent the following clause from standing on its own as a complete sentence. *Watch out* for this kind of construction. It is called a *subordinate clause,* and it is not a sentence.]

Correct: After the latest edition of the newspaper came out, the mayor's press secretary was overwhelmed with phone calls. [A common method of correcting a subordinate clause that has been punctuated as a complete sentence is to connect it to the complete sentence to which it is closest in meaning.]

Incorrect: Several congressmen asked for copies of the Vice President's position paper. Which called for reform of the Environmental Protection Agency.

Correct: Several congressmen asked for copies of the Vice President's position paper, which called for reform of the Environmental Protection Agency.

Spelling

All of us have problems spelling certain words that we have not yet committed to memory. But most writers are not as bad at spelling as they believe themselves to be. Usually it is a handful of words that the individual finds troubling. It is important to be as sensitive as possible to your own particular spelling problems—and to keep a dictionary handy. There is no excuse for failing to check spelling.

Do not rely on your computer's spell checker. There are certain kinds of spelling errors that computers cannot catch, as the following two sentences demonstrate:

- Wilbur wood rather dye than admit that he had been their.
- When he cited the bare behind the would pile, he thought, "Isle just lye hear until he goes buy."

Here are a list of commonly confused words and a list of commonly misspelled words. Read through the lists, looking for those words that tend to give you trouble. If you have any questions, *consult your dictionary.*

Commonly Confused Words

accept/except	affect/effect
advice/advise	aisle/isle

allusion/illusion
an/and
angel/angle
ascent/assent
bare/bear
brake/break
breath/breathe
buy/by
capital/capitol
choose/chose
cite/sight/site
complement/compliment
conscience/conscious
corps/corpse
council/counsel
dairy/diary
descent/dissent
desert/dessert
device/devise
die/dye
dominant/dominate
elicit/illicit
eminent/immanent/imminent
envelop/envelope
every day/everyday
fair/fare
formally/formerly
forth/fourth
hear/here
hole/whole
human/humane
its/it's
know/no
later/latter
lay/lie
lead/led
lessen/lesson
loose/lose

may be/maybe
miner/minor
moral/morale
of/off
passed/past
patience/patients
peace/piece
personal/personnel
plain/plane
precede/proceed
presence/presents
principal/principle
quiet/quite
rain/reign/rein
raise/raze
reality/realty
respectfully/respectively
reverend/reverent
right/rite/write
road/rode
scene/seen
sense/since
stationary/stationery
straight/strait
taught/taut
than/then
their/there/they're
threw/through
too/to/two
track/tract
waist/waste
waive/wave
weak/week
weather/whether
were/where
which/witch
whose/who's
your/you're

Commonly Misspelled Words

a lot
acceptable
accessible
accommodate
accompany

accustomed
acquire
against
annihilate
apparent

arguing
argument
authentic
before
begin
beginning
believe
benefited
bulletin
business
cannot
category
committee
condemn
courteous
definitely
dependent
desperate
develop
different
disappear
disappoint
easily
efficient
environment
equipped
exceed
exercise
existence
experience
fascinate
finally
foresee
forty
fulfill
gauge
guaranteed
guard
harass
hero
heroes
humorous
hurried
hurriedly
hypocrite
ideally
immediately

immense
incredible
innocuous
intercede
interrupt
irrelevant
irresistible
irritate
knowledge
license
likelihood
maintenance
manageable
meanness
mischievous
missile
necessary
nevertheless
no one
noticeable
noticing
nuisance
occasion
occasionally
occurred
occurrences
omission
omit
opinion
opponent
parallel
parole
peaceable
performance
pertain
practical
preparation
probably
process
professor
prominent
pronunciation
psychology
publicly
pursue
pursuing
questionnaire

realize
receipt
received
recession
recommend
referring
religious
remembrance
reminisce
repetition
representative
rhythm
ridiculous
roommate
satellite
scarcity
scenery
science
secede
secession
secretary
senseless
separate
sergeant
shining
significant
sincerely

skiing
stubbornness
studying
succeed
success
successfully
susceptible
suspicious
technical
temporary
tendency
therefore
tragedy
truly
tyranny
unanimous
unconscious
undoubtedly
until
vacuum
valuable
various
vegetable
visible
without
women
writing

Formats

PRELIMINARY CONSIDERATIONS

Your format makes your paper's first impression. Justly or not, accurately or not, it announces your professional competence—or lack of competence. A well-executed format implies that your paper is worth reading. More importantly, however, a proper format brings information to your readers in a familiar form that has the effect of setting their minds at ease. Your paper's format should therefore, impress your readers with your academic competence by following accepted professional standards for criminal justice writing. Like the style and clarity of your writing, your format communicates messages that are often more readily and profoundly received than the content of the document itself.

This chapter contains instructions for the following format elements:

- Margins
- Pagination
- Title page
- Abstract
- Executive summary
- Outline summary
- Table of contents
- List of tables and figures
- Text
- Headings and subheadings
- Tables
- Illustrations and figures
- Reference listing
- Appendixes

There are many format styles available to the criminal justice professional and no consensus as to which one is most appropriate. The majority of journals in the discipline have established their own style. A case in point is the *Justice Quarterly*, the official journal of the Academy of Criminal Justice Sciences, which requires its contributors to use a format comprised of elements from the second edition of the *American Sociological Association Style Guide* (1997) and the fourth edition of the *Publication Manual of the American Psychological Association* (1994). The format described in this chapter is also based on elements from these two comprehensive sources, but it is styled more for the student writer than for the professional.

When preparing a paper for submission to an accredited journal in criminal justice, you should always carefully follow that journal's guidelines for submitting manuscripts; but for now, unless you receive instructions to the contrary from your course instructor, follow the format directions in this manual exactly. Guidelines for citing and referencing sources—using both ASA and APA standards—are explained in Chapter 4.

Criminal justice assignments should be typed or printed on 8 1/2-by-11-inch premium white bond, 20-lb. or heavier. Do not use any other color or size except to comply with special instructions from your instructor, and do not use an off-white or poor quality (draft) paper. *Always submit to your instructor an original typed or computer (preferably laser) printed manuscript. Do not submit a photocopy!* Always make a second copy to keep for your own files and to keep in case the original is lost. If you are using a computer—and we highly recommend that you do—it's a good idea to keep a copy of your paper on the hard drive, another on a disc you can store in a safe place, and a hard copy of the paper in case the computer "crashes" and the disc is lost.

When you have completed your paper and you are ready to turn it in, do not bind it or enclose it within a plastic cover sheet unless instructed to do so. Place one staple in the upper left corner, or use a paper clip at the top of the paper. Note that a paper to be submitted to a journal for publication should not be clipped, stapled, or bound in any form.

MARGINS

Except for theses and dissertations, *margins* should be one inch from all sides of the paper. [*Note:* The *ASA Style Guide* (1997), requires margins to be no less than $1\frac{1}{4}$ inches on all sides for papers submitted for publication.] Unless otherwise instructed, all submissions should be *double-spaced* in a 12-point word-processing font or typewriter pica type (10 characters per inch). Typewriter elite type may be used if pica is not available. Select a font that is plain and easy to read such as Helvetica, Courier, Garamond, or Times Roman. Do not use script, decorative, or elaborate fonts.

PAGINATION

Page numbers should appear in the upper right-hand corner of each page, starting with the second page of the text and continuing consecutively through the paper. No page number should appear on the title page or on the first page of the text. Page numbers should appear one inch from the right side and one-half inch from the top of the page. You should use lowercase roman numerals *(i, ii, iii . . .)* for the front-matter, such as the title page, table of contents, and table of figures that precede the first page of text. These roman numerals should begin with *ii* for the page that immediately follows the title page and centered one-half inch from the bottom of the page.

In special cases, as when your instructor wants no pages preceding the first page of text, the title of the paper should appear one inch from the top of the first page, centered, followed by your name and course information. Most formats omit placing the page number on this first page of text, but some instructors may require it to be placed at the bottom center. All other page numbering should follow the above guidelines.

TITLE PAGE

The following information should be centered on the title page:

- Title of the paper
- Name of writer
- Course name and section number
- Name of instructor
- Name of college or university
- Date

As the sample title page shows, the title should clearly describe the problem addressed in the paper. If the paper discusses juvenile recidivism in Muskogee County jails, for example, the title "Recidivism in the Muskogee County Juvenile Justice System" is professional, clear, and helpful to the reader. "Muskogee County," "Juvenile Justice," or "County Jails" are all too vague to be effective. Also, the title should not be "cute." A cute title may attract attention for a play on Broadway, but it will detract from the credibility of a paper in criminal justice. "Inadequate Solid Waste Disposal Facilities in Muskogee" is professional. "Down in the Dumps" is not.

ABSTRACT

An abstract is a brief summary of a paper written primarily to allow potential readers to see if the paper contains information of sufficient interest for them to read. People conducting research want specific kinds of information, and they often read dozens of abstracts looking for papers that contain relevant data. The heading *Abstract* is centered near the top of the page. Next is the title, also centered, followed by a summary of the paper's topic, research design, results, and conclusions. The abstract should be written in one to three paragraphs, totaling approximately 100 to 150 words. Remember, an abstract is not an introduction; instead, it is a brief summary, as demonstrated in the accompanying sample.

***An Explanation of the Association
Between Crime and Unemployment***

by

George Gonzales

For

Criminology 4473, Section 3

Professor Sid Brown

Middlebury College

October 28, 2001

SAMPLE TITLE PAGE

EXECUTIVE SUMMARY

An executive summary, like an abstract, summarizes the content of a paper but does so in more detail. A sample executive summary is given below. Whereas abstracts are read by people who are doing research, executive summaries are more likely to

ABSTRACT

College Students' Attitudes on Capital Punishment

A survey of college students' beliefs about capital punishment was undertaken in October of 2001 at Ohio State University. The sample was comprised of 92 students in two lower-level criminal justice classes. The purpose of the study was to determine the extent to which students believe it is acceptable for the state to take the life of individuals who commit certain crimes.

Variables tested for association with capital punishment attitudes were age, sex, and ethnicity. Results indicated that the majority of students favor the use of capital punishment with certain crimes, and that young white males are the most pro-capital punishment. The students in this sample expressed a concern about what might happen to the rate of serious crimes if the death penalty were abolished. However, they were more concerned with those who committed certain heinous crimes being punished appropriately.

be read by people who need some or all of the information in the paper in order to make a decision. Many people, however, will read the executive summary to fix clearly in their mind the organization and results of a paper before reading the paper itself.

The length of the executive summary is dictated to some extent by the length of the document being summarized. For example, a 150 to 200-page grant report might be summarized in an 8 to 10-page executive summary, while a 25-page research paper might need only a page or two of summary. The lengthier executive summary might also contain headings and subheadings similar to those used in the actual report. By following the outline of the actual paper, the executive summary serves as a sort of summary of each of the sections in the much lengthier report.

The executive summary on the next page summarizes a report on the second year of a *Drug Assessment Study with Juvenile Offenders in a Secure Detention Setting* (Johnson, Rettig and Steward 1993). This example gives you an idea of the details that should be included in your summary.

EXECUTIVE SUMMARY

This *Drug Assessment Study* is written for practitioners who deal with drug-involved juvenile offenders repeatedly cycling through detention centers. The assessment process was initiated sometime after they were placed in detention. Two psychologists interviewed 198 youth over a six-month period and recorded the information they obtained on a specially designed "assessment" instrument. They also administered an instrument designed to measure the level of alcohol and other drug use, coupled with an unobtrusive measurement of denial and other underlying manifestations of substance abuse. The results of this study are summarized below.

One important finding had to do with interpersonal relationships. Those who used and abused alcohol and/or other drugs were generally friends of users and abusers. Those who did not use and abuse these substances almost always reported much less association with those who use and abuse drugs. Young people on probation and parole for substance abuse–related offenses probably should be restrained by the court from association with those who have a serious alcohol and/or drug history.

The findings from this study support the hypothesis that where youth are distanced from the family, in terms of communication, feelings, and dis-organized relationships, their chances of alcohol and/or other drug in-volvement increase, at times dramatically. Significant progress in helping delinquents change and adjust might be attained by a serious commitment to family therapy and family development, especially with the cooperation and insistence of the court. There is good reason to believe that significant differences exist between white and minority families with respect to child-rearing philosophy and practice. If this is true, it may well affect alcohol and/or other drug use, attitudes toward families, school, and other areas of juvenile misbehavior. It would follow that detention treatment plans, court-ordered probation and aftercare alternatives should be differentiated, where feasible and legal, to respond to minority families at the point of their needs.

Findings from this study imply that for those youth who come in contact with the juvenile justice system: (1) blacks are much less addicted than either whites or other minorities, at least in very early adolescence;

(2) black parents are significantly more concerned with their children's use of alcohol and/or other drugs when compared with white parents; (3) blacks report much more emotional support from their family members, and (4) there appears to be, at least on the surface, more dysfunctional family interaction among white families than black. Perhaps the juvenile justice system should develop neighborhood programs to provide positive support within the strong kinship networks already in place in the black culture; early intervention could reach black children *before* they fully enter the rebellious youth culture.

The body of literature cited in this report is fairly unified in asserting that common etiological roots cannot be shown between substance abuse and delinquency. Even the relationship between violent crime and substance abuse remains clouded (Inciardi 1981). A relationship between substance use and more serious delinquency appears to be developmental rather than causal (Huizinga et al. 1989). More than ever, this suggests that attention should be given to the National Council of Juvenile and Family Court Judges' report, which argues from a systemic and holistic perspective (*Criminal Justice Newsletter* 1986). Drug use and abuse, child neglect, abandonment, sexual, physical and emotional abuse, family violence, family dysfunction, and juvenile delinquency are interactive variables that cannot be clearly identified, diagnosed, or treated without addressing them together.

Our study supports a vast body of literature that suggests meaningful intervention in the lives of juveniles at risk should include, whenever possible, a holistic approach. Not only must each youth be requested to take responsibility for his or her behavior and work toward resolution of his problems, but also family members should be called to account for their responsibility and compelled, when necessary, to participate in treatment.

OUTLINE SUMMARY

An outline page is a specific style of executive summary. It clearly shows the sections into which the paper is divided and the content of the information in each section. An outline summary is an asset to busy decision makers because it allows them to under-

---

OUTLINE SUMMARY

I. *The problem* is that picnic and restroom facilities at Hafer Community Park have deteriorated.

   A. Only one major renovation has occurred since 1967, when the park was opened.

   B. The Park Department estimates that repairs would cost about $33,700.

II. Three possible solutions have been given extensive consideration:

   A. One option is to do nothing. Area residents will use the park less as deterioration continues.

   B. The first alternative solution is to make all repairs immediately, which will require a total of $33,700.

   C. A second alternative is to make repairs according to a priority list over a five-year period.

III. The recommendation of this report is that alternative C be adopted by the city council. The benefit/cost analysis demonstrates that residents will be satisfied if basic improvements are made immediately.

---

stand the entire content of a paper without reading it, or to refer quickly to a specific part of the paper for more information.

TABLE OF CONTENTS

A table of contents does not provide as much information as an outline but does include the headings of the major sections and subsections of a paper. Tables of contents are not normally required in student papers or papers presented at professional meetings, but they may be included. (The *ASA Style Guide* does not require a table of contents for manuscripts submitted for publication consideration.) Tables of contents are normally required, however, in books, theses, and dissertations. The table of contents should consist of the chapter or main section titles and the first level of headings within sections, along with their page numbers, as the accompanying example demonstrates.

TABLE OF CONTENTS

EXECUTIVE SUMMARY	iv
INTRODUCTION	1
REVIEW OF LITERATURE	5
DESIGN	26
RESULTS	28
Alcohol and Other Drugs	30
Sex	33
Race	34
School and Substance Abuse	39
Suicide	40
DISCUSSION	41
REFERENCES	52
APPENDIX 1: Assessment	59
APPENDIX 2: T-Tests	66
APPENDIX 3: Analyses of Variance	80
APPENDIX 4: Cross-Tabulations	99
APPENDIX 5: Frequencies and Percentages	123
APPENDIX 6: Program Evaluation	132
APPENDIX 7: Project Workplan/Objectives/Evaluation	139

LISTS OF TABLES AND FIGURES

A list of tables or figures contains the titles of the tables or figures included in the paper in the order in which they appear, along with their page numbers. You may list tables, illustrations, and figures together under the title "Figures" (and call them all figures in the text), or if you have a list with more than a half-page of entries, you may have separate lists of tables, figures, and illustrations (and title them accordingly in the text). The format for all such tables should follow the accompanying example.

LIST OF FIGURES

1. Population Growth in Five U.S. Cities, 1980–1986   1
2. Welfare Recipients by State, 1980 and 1990   3
3. Economic Indicators, January to June 1991   6
4. Educational Reforms at the Tennessee Convention   11
5. International Trade, 1880–1890   21
6. Gross Domestic Product, Nova Scotia, 1900–1960   22
7. California Domestic Program Expenditures, 1960–1980   35
8. Juvenile Recidivism in Illinois   37
9. Albuquerque Arts and Humanities Expenditures, 1978   39
10. Railroad Retirement Payments after World War II   40

TEXT

Ask your instructor for the number of pages required for the paper you are writing. The text should follow the directions explained in chapters 1 and 2. Headings and subheadings within the text vary with the assignment. Procedures for applying headings and subheadings to your text are described below.

HEADINGS AND SUBHEADINGS

Generally, three heading levels should meet your organizational needs when writing criminal justice papers:

1. *Primary headings* should be centered and printed in all capital letters.
2. *Secondary headings* should be centered and italicized, with only the first letter in each word capitalized, excluding articles, prepositions, and conjunctions.
3. *Tertiary headings* should be indented and lead into the paragraph with only the first letter of the first word capitalized. They should be italicized and followed by a period. The text should follow.

The accompanying example has three heading levels that follow these guidelines.

> **RESULTS**
> *Alcohol and Other Drugs*
> *Risk factors.* Characteristics of risk can be found in families where adolescents . . .

SAMPLE: THREE LEVELS OF HEADINGS

TABLES

Tables are used in the text to show relationships among data to help the reader come to a conclusion or understand a certain point. Tables that show simple results or "raw" data should be placed in an appendix. Tables should not reiterate the content of the text. They should say something new, and they should stand on their own. That is, the reader should be able to understand the table without reading the text. Clearly label the columns and rows in each table. Each word in the title (except articles, prepositions, and conjunctions) should be capitalized. The source of the information should be shown immediately below the table, not in a footnote or endnote. A hypothetical table on population change is shown as an example.

TABLE 1 *Population Change in Ten U.S. Cities 1980–1986*

City	1986 Rank	1980 Population	1986 Population	Percentage Change 1980 to 1986
New York	1	7,071,639	7,262,700	2.7
Los Angeles	2	2,968,528	3,259,300	9.8
Chicago	3	3,005,072	3,009,530	0.2
Houston	4	1,611,382	1,728,910	7.3
Philadelphia	5	1,688,210	1,642,900	−2.3
Detroit	6	1,203,369	1,086,220	−9.7
San Diego	7	875,538	1,015,190	16.0
Dallas	8	904,599	1,003,520	10.9
San Antonio	9	810,353	914,350	12.8
Phoenix	10	790,183	894,070	13.1

Source: U.S. Bureau of the Census (1988).

ILLUSTRATIONS AND FIGURES

Illustrations are not normally inserted in the text of a criminal justice paper, and they are not included even in the appendix unless they are necessary to explain the material in the text. If illustrations are necessary, do not paste or tape photocopies of photographs or similar materials to the text or the appendix. Instead, photocopy each one on a separate sheet of paper and center it, along with its typed title, within the normal margins of the paper. The format of the illustration titles should be the same as that for tables and figures.

Figures in the form of charts and graphs may be very helpful in presenting certain types of information. The example below demonstrates how data in the preceding table can be presented in a bar chart.

REFERENCE LISTING

Chapter 4 discusses and gives models of two standard citation systems for the references in your paper: the ASA style and the APA style. Some instructors prefer papers to be structured in *article format,* with everything presented as tightly compressed and succinct as possible. If your instructor favors this system, your reference section should immediately follow (double-spaced) the last line of your discussion section. Other instructors prefer the references to be listed on a separate page. Ask your instructor which system you should follow.

APPENDIXES

Appendixes are reference materials for the convenience of the reader at the back of the paper, after the text. Providing information that supplements the important facts contained in the text, they may include maps, charts, tables, and selected documents.

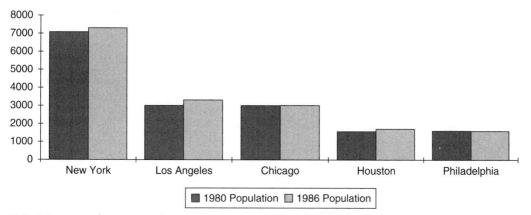
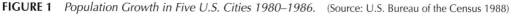

FIGURE 1   *Population Growth in Five U.S. Cities 1980–1986.*   (Source: U.S. Bureau of the Census 1988)

Do not place materials that are merely interesting or decorative in your appendix. Include only items that will answer questions raised by the text or that are necessary to explain the text.

Follow the format guidelines for illustrations, tables, and figures when adding material in an appendix. At the top center of the page, label your first appendix "Appendix A," your second appendix "Appendix B," and so on. If you are only appending one item, it should be labeled "Appendix," with no letter indicating sequence. Do not append an entire government report, journal article, or other publication, but only the portions of such documents that are necessary to support your paper. The source of the information should always be evident on the appended pages.

4

Citing Sources

PRELIMINARY DECISIONS

One of your most important jobs as a research writer is to document your use of source material carefully and clearly. Failure to do so will cause your reader confusion, damage the effectiveness of your paper, and perhaps make you vulnerable to a charge of plagiarism. Proper documentation is more than just good form. It is a powerful indicator of your own commitment to scholarship and the sense of authority that you bring to your writing. Good documentation demonstrates your expertise as a researcher and increases your reader's trust in you and your work.

Unfortunately, as anybody who has ever written a research paper knows, getting the documentation right can be a frustrating, confusing job, especially for the novice writer. Positioning each element of a single reference citation accurately can require what seems an inordinate amount of time spent thumbing through the style manual. Even before you begin to work on specific citations, there are important questions of style and format to answer.

What to Document

Direct quotes must always be credited, as must certain kinds of paraphrased material. Information that is basic—important dates, facts, or opinions universally acknowledged—need not be cited. Information that is not widely known, whether fact or opinion, should receive documentation.

What if you are unsure whether or not a certain fact is widely known? You are, after all, very probably a newcomer to the field in which you are conducting your research. If in doubt, supply the documentation. It is better to overdocument than to fail to do justice to a source.

The Choice of Style

While the question of which documentation style to use may be decided for you in some classes by your instructor, others may allow you a choice. There are several styles available, each designed to meet the needs of writers in particular fields. As mentioned earlier, many journals within the field of criminal justice establish their own documentation styles. Perhaps the two styles most widely accepted within the discipline are those of the American Sociological Association, published in the second edition of the *ASA Style Guide* (1997), and the American Psychological Association (APA), published in the fourth edition of the *Publication Manual of the American Psychological Association* (1994). This chapter will describe both styles in sufficient detail for you to use either one.

Read through the following pages before trying to use them to structure your notes. Unpracticed student researchers tend to ignore the documentation section of their style manual until the moment the first note has to be worked out, and then they skim through the examples looking for the one that perfectly corresponds to the immediate case in hand. But most style manuals do not include every possible documentation model, so the writer must piece together a coherent reference out of elements from several models. Reading through all the models before using them gives you a feel for where to find different aspects of models as well as for how the referencing system works in general.

CITING SOURCES IN ASA STYLE

The American Sociological Association (ASA) system of style is widely used by students and professionals. The ASA has adopted a modification of the style elaborated in the *Chicago Manual of Style* (1993)—hereafter *CMS*—perhaps the most universally approved of all documentation authorities. Throughout this section, where answers to questions of style are not clearly defined and explained in the *ASA Style Guide* (1997), we have referred to the *CMS* for recommendations.

One of the advantages of using the ASA style, which is outlined in a pamphlet entitled *ASA Style Guide* (1997), is that it is designed to guide professionals in preparing a manuscript for submission to a journal. The ASA style is required for all papers submitted to the *American Sociological Review*, the official journal of the ASA and the most influential sociology journal in publication. It is also required for all the leading journals in sociology and some in criminal justice.

A parenthetical reference or citation is a note placed within the text, near where the source material occurs. In order not to distract the reader from the argument, the citation is as brief as possible, containing just enough information to refer the reader to the full reference listing that appears in the bibliography or reference section following the text. Usually the minimum information necessary is the author's last name—meaning the name by which the source is alphabetized in the references at the end of the paper—and the year of the publication of the source. As indicated by the following models, this information can be given in a number of ways. Models of reference listing bibliographical entries that correspond to these parenthetical citations are given in the next section of this chapter.

Text Citations in ASA Style

Citations within the text should include the author's last name and the year of publication. Page numbers should be included only when quoting directly from a source or referring to specific passages. Subsequent citations of the same source should be identified the same way as the first (American Sociological Association 1997). The following examples identify the *ASA Style Guide*'s (1997) citation system for a variety of possibilities.

Author's Name in Text

When the author's name is in the text, it should be followed by the publication year in parentheses:

> Freedman (1984) postulates that when individuals . . .

Author's Name Not in Text

When the author's name is not in the text, the last name and publication should be enclosed in parentheses:

> . . . encourage more aggressive play (Perrez 1979).

Citation including Page Numbers

Page numbers are included only when quoting directly from a source or referring to specific passages. However, some instructors prefer page numbers for all citations in order to check for plagiarism. Ask your instructor what system you should follow. When the page number is included, it should follow the publication year and be preceded by a colon with no space between the colon and the page number:

> Thomas (1961:741) builds on this scenario . . .

Source with Two Authors

When the publication has two authors, both last names should be given:

> . . . establish a sense of self (Holmes and Watson 1872:114–116).

Source with Three Authors

When a publication has three authors, all three last names should be cited in the first citation, with et al. used for subsequent citations in the text. In first citation,

> . . . found the requirements very restrictive (Mollar, Querley, and McLarry 1926).

thereafter,

> . . . proved to be quite difficult (Mollar et al. 1926).

Source with More Than Three Authors

For more than three authors, use et al. in all citations.

> Kinneson et al. (1933) made the . . .

Two Authors with the Same Last Name

When citing two authors with the same last name, use a first initial to differentiate between them.

> . . . the new budget cuts (K. Grady 1994).
>
> . . . stimulate economic growth (B. Grady 1993).

Two Works by the Same Author

When citing two works by the same author, in the same note, place a comma between the publication dates of the works.

> George (1992, 1994) argues for . . .

If the two works were published in the same year, differentiate between them by adding lowercase letters to the publication dates. Be sure to add the letters to the references in the bibliography, too.

> . . . the city government (Estrada 1994a, 1994b).

Work with No Author Given

According to the CMS (16.40), if the author's name is not provided within the work but you know the author's identity, you may give the name in brackets:

> . . . cannot be held accountable ([Logan] 1994).

If you cannot ascertain the author's name, the CMS (16.41) suggests you begin the citation with the work's title, followed by the date of publication. Do not use the phrase *Anonymous* or *Anon.*

> . . . a logical starting point (*Saving the Underclass* 1985).

Direct Quotes

Direct quotes of fewer than four lines should be placed in the text with quotation marks at the beginning and end. The citation should include the page number in one of the following formats:

> The majority of these ads promote the notion that "If you are slim, you will also be beautiful and sexually desirable" (Rockett and McMinn 1990:278).
>
> Smith and Hill (1997) found that "women are far more likely to obsess about weight" (p. 127).

Direct quotes of four lines or more should be indented from both sides, single spaced, and presented in a smaller font or type when possible. They should be blocked—no tab set for the first line—with no quotation marks as follows:

> According to Brown (1985):

> There are few girls and women of any age or culture raised in white America, who do not have some manifestation of the concerns discussed here, i.e., distortion of body image, a sense of "out-of-control" in relationship to food, addiction to dieting, binging, or self-starvation. (P. 61)

It should be noted that in the block quote the author, date, and or page number follows the period at the end, and that the *P* for "page" is capitalized when the page number appears alone without the author and date, as in this example.

Source Cited in a Secondary Source

Sometimes information is obtained from *a source that is cited in a secondary source.* While it is always best to locate and cite the original source, sometimes this is not possible. When citing a source that is itself cited in a secondary source, refer in your parenthetical citation to the original source, and not the later source in which the original is cited. For example, if you wish to cite information from a 1999 article by John Smith which you have found cited in a 2000 article by Allen Jones, your citation will look like this:

> . . . the promise of a subsequent generation (Smith 1999).

See *Article Cited in a Secondary Source* in the reference section below for information on how to list this citation in your references.

Chapters, Tables, Appendixes, etc.

Chapters, tables, appendixes, etc., should be cited as follows:

> . . . (Johnson 1995, chap. 6).
> . . . (Blake 2001, table 4:34).
> . . . (Cleary 1976, appendix C:177).

Reprints

When citing a work reprinted from an earlier publication, give the earliest date of publication in brackets, followed immediately by the date of version you have used:

> . . . Baldwin ([1897] 1992) interpreted this . . .

More Than One Source in a Reference

When citing more than one source, separate citations by a semicolon and order them in a manner of your choice. You may arrange them in alphabetical order, date order, or order of importance to your argument, but whatever order you choose, use it consistently throughout your paper.

> . . . are related (Harmatz 1987:48; Marble et al. 1986:909; Powers and Erickson 1986:48; Rackley et al. 1988:10; Thompson and Thompson 1986:1067).

Unpublished and/or Undated Materials

The date should be given for dissertation and unpublished papers. When the date is not available, use *n.d.* (no date) in place of the date. Use the word *forthcoming* when materials cited are unpublished but scheduled for publication.

> Studies by Barkely (forthcoming) and Jorden (n.d.) lend support . . .

Archival Sources

When citing National Archives or other archival sources, abbreviate citations:

> . . . (NA, 266, Box 567, June 15, 1952).

Machine Readable Data Files

When citing machine-readable data files, authorship and date should be included:

> . . . (American Institute of Public Opinion 1989).

Classic Texts

When citing classic texts, such as the Bible, standard translations of ancient Greek texts, or numbers of the *Federalist Papers,* you may use the systems by which they are subdivided. Since any edition of a classic text employs the standard subdivisions, this reference method has the advantage of allowing your reader to find the source passage in any published edition of the text. It is not necessary to include a citation for a classic text in the reference section.

You may cite a biblical passage by referring to the particular book, chapter, and verse, all in roman type, with the translation given after the verse number:

> "But the path of the just is as the shining light, that shineth more and more unto the perfect day" (Proverbs 4:18 King James Version).

The Federalist Papers are numbered:

> Madison addresses the problem of factions in a republic (*Federalist* 10).

Newspapers

According to the *CMS* (16.117), references to material in daily newspapers should be handled within the syntax of your sentence:

> In an August 10, 1993, editorial, the *New York Times* painted the new regime in glowing colors.
>
> An article entitled "Abuse in Metropolis," written by Harry Black and published in the *Daily Planet* on December 24, 1996, took exception to Superman's remarks.

According to the *CMS* (16.117) references to newspaper items are not usually included in the reference list or bibliography. If you wish to include newspaper references, however, there is a model of a bibliographical entry in the next section of this chapter.

Public Documents

When citing a public/government document or one with institutional authorship, you should supply the minimum identification:

. . . (U.S. Bureau of the Census 2002:223).

Since the *ASA Style Guide* gives formats for only two types of government publications, the following models are based not only on practices from the ASA guide but also on formats found in the *CMS* (15.322–411, 16.148–79). Corresponding bibliography entries appear in the next section.

Parenthetical text references to both the *Senate Journal* and the *House Journal* start with the journal title in place of the author, the session year, and, if applicable, the page:

(*Senate Journal* 1997:24).

Congressional debates are printed in the daily issues of the *Congressional Record,* which are bound biweekly and then collected and bound at the end of the session. Whenever possible, you should consult the bound yearly collection instead of the biweekly compilations. Your parenthetical reference should begin with the title *Congressional Record* (or *Cong. Rec.*) in place of the author's name and include the year of the congressional session, the volume and part of the *Congressional Record,* and finally the page:

(*Cong. Rec.* 1930, 72, pt. 8:9012).

References to congressional reports and documents, which are numbered sequentially in one- or two-year periods, include the name of the body generating the material, the year, and the page:

(U.S. House 2001:12).

NOTE: any reference that begins with U.S. Senate or U.S. House may omit the U.S., if it is clear from the context that you are referring to the United States. Whichever form you use, be sure to use it consistently, in both the notes and the bibliography.

Bills and Resolutions

According to the *CMS* (15.347–48), bills and resolutions, which are published in pamphlets called "slip bills," on microfiche, and in the *Congressional Record,* are not always given a parenthetical text reference and a corresponding bibliography entry. Instead, the pertinent reference information appears in the syntax of the sentence. If, however, you wish to cite such information in a text reference, the form depends on the source from which you took your information. If citing to a slip bill, use one of these forms:

(U.S. Senate 1996)

(Visa Formalization Act of 1996)

You may cite either the body that authored the bill or the title of the work itself. Whichever method you choose, remember to begin your bibliography entry with the same material. Here is a model for citing to the *Congressional Record:*

(U.S. Senate 1996:S7658).

The number following the date and preceded by an S (for Senate; H for House) is the page in the *Congressional Record.*

As with bills and resolutions, laws (also called statutes) are not necessarily given a parenthetical text reference and a bibliography entry. Instead, the identifying material is included in the text. If you wish to make a formal reference for a statute, you must structure it according to the place where you found the law published. Initially published separately in pamphlets, as slip laws, statutes are eventually collected and incorporated, first into a set of volumes called *U.S. Statutes at Large* and later into the *United States Code,* a multivolume set that is revised every six years. You should use the latest publication. When citing a slip law, you should use either *U.S. Public Law,* in roman type, and the number of the piece of legislation, or the title of the law:

(U.S. Public Law 678:16–17)

(Library of Congress Book Preservation Act of 1997:16–17)

When citing to the *Statutes at Large,* use this form:

(Statutes at Large 2001:466)

The following form is for citing to the *United States Code:*

(Library of Congress Book Preservation Act of 1997, U.S. Code. Vol. 38, Sec. 1562)

United States Constitution

According to the *CMS* (15.367), references to the United States Constitution include the number of the article or amendment, the section number, and the clause, if necessary:

(U.S. Constitution, art. 3, sec. 3)

It is not necessary to include the Constitution in the bibliography.

Executive Department Documents

A reference to a report, bulletin, circular, or any other type of material issued by the Executive Department starts with the name of the agency issuing the document, although you may use the name of the author, if known:

(Department of Labor 1984:334).

Legal References

United States Supreme Court

As with laws, Supreme Court decisions are rarely given their own parenthetical text reference and bibliography entry but are instead identified in the text. If you wish to use a formal reference, however, you may place within the parentheses the title of the case, in italics, followed by the source (for cases after 1875 this is the *United States Supreme Court Reports,* abbreviated U.S.), which is preceded by the volume number and followed by the page number. You should end the first reference to the case that appears in your paper with the date of the case, in brackets. You need not include the date in subsequent references:

(*State of Nevada v. Goldie Warren* 324 U.S. 123 [1969])

Before 1875, Supreme Court decisions were published under the names of official court reporters. The reference below is to William Cranch, *Reports of Cases Argued and Adjudged in the Supreme Court of the United States, 1801–1815*, 9 vols. (Washington, D.C., 1804–17). The number preceding the clerk's name is the volume number; the last number is the page:

. . . in that now famous case (1 Cranch 137)

For most of these parenthetical references, it is possible to move some or all of the material outside the parentheses simply by incorporating it in the text:

In 1969, *in State of Nevada v. Goldie Warren* (324 U.S. 123), the judge ruled that . . .

Lower Courts

Decisions of lower federal courts are published in the *Federal Reporter.* The note should give the volume of the *Federal Reporter (F.),* the series, if it is other than the first (*2d,* in the model below), the page, and in brackets, an abbreviated reference to the specific court (the example below is to the Second Circuit Court) and the year:

(*United States v. Sizemore,* 183 F. 2d 201 [2d Cir. 1950])

Publications of Government Commissions

According to the *CMS* (15.368) references to bulletins, circulars, reports, and study papers that are issued by various government commissions should include the name of the commission, the date of the document, and the page:

(Securities and Exchange Commission 1985:57)

Corporate Authors

Because government documents are often credited to a corporate author with a lengthy name, you may devise an acronym or a shortened form of the name and indicate in your first reference to the source that this name will be used in later citations:

(Bulletin of Labor Statistics 1997, 1954; hereafter *BLS)*

The practice of using a shortened name in subsequent references to any corporate author, whether a public or private organization, is sanctioned in most journals and approved in the *CMS* (15.252). Thus, if you refer often to the *U.N. Monthly Bulletin of Statistics,* you may, after giving the publication's full name in the first reference, use a shortened form of the title—perhaps *UNMBS*—in all later cites.

Publications of State and Local Governments

According to the *CMS* (15.377), references to state and local government documents are similar to those for the corresponding national government sources:

(Oklahoma Legislature 1995:24)

The *CMS* (16.178), restricts bibliographical information concerning state laws or municipal ordinances to the running text.

Interviews

According to the *CMS* (16.127, 16.130), in the author-date system, citations to interviews should be handled by references within the text—in the syntax of a sentence—rather than in parentheses:

In a March 1997 interview with O. J. Simpson, Barbara Walters asked questions that seemed to upset and disorient the former superstar.

For published or broadcast interviews, no parenthetical reference is necessary, but there should be a complete citation under Walter's name in the bibliography.

An unpublished interview conducted by the writer of the paper should also be cited in the syntax of the sentence:

In an interview with the author on April 23, 1993, Dr. Kennedy expressed her disappointment with the new court ruling.

If you are citing material from an interview that you conducted, identify yourself as "the author" and give the date of the interview. Cite the interview by placing the date in parentheses following the name of the person interviewed:

Marsha Cummings (2000), Director of the Children's Hospital in Oklahoma City, was interviewed by the author on November 14, 2000.

References in ASA Style

Parenthetical citations in the text point the reader to the fuller source descriptions at the end of the paper known as the references, or bibliography. This reference list, which always directly follows the text under the heading *REFERENCES* (this is considred a primary heading in the paper you are writing and follows the guidelines of headings listed on pages 57 and 58), is arranged alphabetically according to the first element in each citation. As stated in Chapter 3, some instructors prefer papers to be structured in *article format,* with everything presented as tightly compressed and succinct as possible. If your instructor favors this system, your reference section should immediately follow (double space) the last line of your discussion section. Other instructors prefer the references to be listed on a separate page. Ask your instructor which system you should follow.

As with most alphabetically arranged bibliographies, there is a kind of reverse-indentation system: After the first line of a citation, all subsequent lines are indented five spaces. The entire reference section is double-spaced.

The ASA uses standard, or "headline style," capitalization for titles in the reference list. In this style, all first and last words in a title, and all other words except articles *(a, an, the)*, coordinating words *(and, but, or, for, nor)*, and prepositions *(among, by, for, of, to, toward,)* are capitalized.

Remember that every source cited in the text, with those exceptions noted in the examples below, must have a corresponding entry in the references section. *Do not include references to any work not cited in the text of your paper.*

Most of the following formats are based on those given in the *ASA Style Guide* (1997). Formats for bibliographical situations not covered by the ASA guide are taken from the *Chicago Manual of Style* (1993).

BOOKS

One Author

First comes the author's name, inverted, then the date of publication, followed by the title of the book, the place of publication, and the name of the publisher. Use first names for all authors or initials if no first name is provided. Add a space after each initial, as in the example below. For place of publication, always identify the state unless the city is New York. Use postal abbreviations to denote states (OK, MA, etc.).

Periods are used to divide most of the elements in the citation, although a colon is used between the place of publication and publisher. Custom dictates that the main title and subtitle be separated by a colon, even though a colon may not appear in the title as printed on the title page of the book.

> Northrup, A. K. 1997. *Living High off the Hog: Recent Pork Barrel Legislation in the Senate.* Cleveland, OH: Johnstown.

Two Authors

Only the name of the first author is reversed, since it is the one by which the citation is alphabetized. Note that there is no comma between the first name of the first author and the *and* following:

> Spence, Michelle and Kristen Ruell. 1996. *Hiring and the Law.* Boston, MA: Tildale.

Three or More Authors

The use of *et al.* is not acceptable in the references section; list the names of all authors of a source. While the ASA style places commas between all names in the text citation—(Moore, Rice, and Traylor 1998)—it deletes the comma separating the next-to-last and last names in the bibliographical reference. Note also that the ASA does not advocate abbreviating the word *University* in the name of a press, as indicated in the model below.

> Moore, J. B., Allen Rice and Natasha Traylor. 1998. *Down on the Farm: Culture and Folkways*. Norman, OK: University of Oklahoma Press.

Work with No Author Given

The *CMS* (16.40) states that if you can ascertain the name of the author when that name is not given in the work itself, place the author's name in brackets:

> [Morey, Cynthia]. 1977. *How We Mate: American Dating Customs, 1900–1955.* New York: Putney.

Do not use the phrase "anonymous" to designate an author whose name cannot be determined; instead, according to the *CMS* (16.41), begin your reference entry with the title of the book, followed by the date. You may move initial articles *(a, an, the)* to the end of the title:

> *Worst Way to Learn: The Government's War on Education, The.* 1997. San Luis Obispo, CA: Blakeside.

Editor, Compiler, or Translator as Author

When no author is listed on the title page, begin the citation with the name of the editor, compiler, or translator:

> Trakas, Dylan, comp. 1998. *Making the Road-Ways Safe: Essays on Highway Preservation and Funding.* El Paso, TX: Del Norte Press.

Editor, Compiler, or Translator with Author

> Pound, Ezra. 1953. *Literary Essays.* Edited by T. S. Eliot. New York: New Directions.
> Stomper, Jean. 1973. *Grapes and Rain.* Translated by John Picard. New York: Baldock.

Untranslated Book

If your source is in a foreign language, it is not necessary, according to the *CMS* (15.118), to translate the title into English. Use the capitalization format of the original language.

> Picon-Salas, Mariano. 1950. *De la Conquista a la Independencia.* Mexico D. F.: Fondo de Cultura Económica.

If you wish to provide a translation of the title, do so in brackets or parentheses following the title. Set the translation in roman type, and capitalize only the first word of the title and subtitle, proper nouns, and proper adjectives:

> Wharton, Edith. 1916. *Voyages au front* (Visits to the front). Paris: Plon.

Two or More Works by the Same Author

The author's name in all citations after the first may be replaced, if you wish, by a three-em dash (six strokes of the hyphen):

> Russell, Henry. 1978. *Famous Last Words: Notable Supreme Court Cases of the Last Five Years*. New Orleans, LA: Liberty Publications.
>
> ———. 1988. *Great Court Battles*. Denver, CO: Axel and Myers.

Chapter in a Multiauthor Collection

> Gray, Alexa North. 1998. "Foreign Policy and the Foreign Press." Pp. 188–204 in *Current Media Issues*, edited by Barbara Bonnard and Luke F. Guinness. New York: Boulanger.

The parenthetical text reference may include the page reference:

> (Gray 1998:195–97)

You must repeat the name if the author and the editor are the same person:

> Farmer, Susan A. 1995. "Tax Shelters in the New Dispensation: How to Save Your Income." Pp. 58–73 in *Making Ends Meet: Strategies for the Nineties*, edited by Susan A. Farmer. Nashville, TN: Burkette and Hyde.

Author of a Foreword or Introduction

According to the *CMS* (16.51), there is no need to cite the author of a foreword or introduction in your bibliography, unless you have used material from that author's contribution to the volume. In that case, the bibliography entry is listed under the name of the author of the foreword or introduction. Place the name of the author of the work itself after the title of the work:

> Farris, Carla. 1998. Foreword to *Marital Stress among the Professoriat: A Case Study,* by Basil Givan. New York: Galapagos.

The parenthetical text reference cites the name of the author of the foreword or introduction, not the author of the book:

> (Farris 1998)

Selection in a Multiauthor Collection

> Gray, Alexa North. 1998 "Foreign Policy and the Foreign Press." Pp. 188–204 in *Current Media Issues,* edited by Barbara Bonnard and Luke F. Guinness. New York: Boulanger.

The parenthetical text reference may include the page reference:

> . . . (Gray 1998:195–97)

You must repeat the name if the author and the editor are the same person:

> Farmer, Susan A. 1995. "Tax Shelters in the New Dispensation: How to Save Your Income." Pp.58–73 in *Making Ends Meet: Strategies for the Nineties,* edited by Susan A. Farmer. Nashville, TN: Burkette and Hyde.

Articles in Encyclopedias and Other Reference Books

According to the *CMS* (15.293), well-known reference books such as the *Encyclopedia Britannica* or the *American Heritage Dictionary* are not given a citation in the references list. They should be credited within the running text of the paper:

> . . . in Rondal Gould's article on welfare in the twelfth edition of *Collier's Encyclopedia . . .*

Subsequent Editions

If you are using an edition of a book other than the first, you must cite the number of the edition or the status, such as *Rev. ed.* for revised edition, if there is no edition number:

> Hales, Sarah. 2000. *The Coming Water Wars*. 3d ed. Pittsburgh, PA: Blue Skies.

Multivolume Work

If you are citing a multivolume work in its entirety, use the following format:

> Graybosch, Charles. 1988–89. *The Rise of the Unions*. 3 vols. New York: Starkfield.

If you are citing only one of the volumes in a multivolume work, use the following format:

> Graybosch, Charles. 1988. *The Beginnings*. Vol. 1 of *The Rise of the Unions*. New York: Starkfield.

Reprints

> Adams, Sterling R. [1964] 2001. *How to Win an Election: Promotional Campaign Strategies*. New York: Starkfield.

Classic Texts

According to the *CMS* (15.294, 15.298), references to classic texts such as sacred books and Greek verse and drama are usually confined to the text and not given citations in the bibliography.

PERIODICALS

JOURNAL ARTICLES. Journals are periodicals, usually published either monthly or quarterly, that specialize in serious scholarly articles in a particular field.

Journal with Continuous Pagination

Most journals are paginated so that each issue of a volume continues the numbering of the previous issue. The reason for such pagination is that most journals are bound in libraries as complete volumes of several issues; continuous pagination makes it easier to consult these large compilations:

> Hunzecker, Joan. 1987. "Teaching the Toadies: Cronyism in Municipal Politics." *Review of Local Politics* 4:250–62.

> Johnson, J. D., N. E. Noel and J. Sutter-Hernandez. 2000. "Alcohol and Male Acceptance of Sexual Aggression: The Role of Perceptual Ambiguity." *Journal of Applied Social Psychology* 30(6):1186–1200.

Note that the name of the journal, which is italicized, is followed without punctuation by the volume number, which is itself followed by a colon and the page numbers. There should be no space between the colon and the page numbers, which are inclusive. Do not use *p.* or *pp.* to introduce the page numbers.

Journal in Which Each Issue Is Paginated Separately

The issue number appears in parentheses immediately following the volume number.

> Skylock, Browning. 1991. "'Fifty-Four Forty or Fight!': Sloganeering in Early America." *American History Digest* 28(3):25–34.

> Entwisle, Doris, Karl Alexander and Linda Olson. 2000. "Urban Youth, Jobs, and High School." *American Sociological Review* 65(2):279–297.

Article Published in More than One Journal Issue

> Crossitch, Vernelle. 1997. "Evaluating Evidence: Calibrating Ephemeral Phenomena," parts 1–4. *Epiphanic Review* 15:22–29; 16:46–58; 17:48–60.

Articles Published in Foreign-Language Journals

Sczaflarski, Richard. 2001 "The Trumpeter in the Tower: Solidarity and Legend" (in Polish). *World Political Review* 32:79–95.

Article Cited in a Secondary Source

When referencing a source that has itself been cited in a secondary source, first list the complete citation of the source you cited, followed by *cited in,* and a listing of the source from which you obtained your citation.

Johnson, William A. and Richard P. Rettig. 1995. "Drug Assessment of Juveniles in Detention." *Social Forces* 28(3):56–69, cited in John Duncan and Mary Ann Hopkins. 2000. "Youth and Drug Involvement: Families at Risk." *British Journal of Addiction* 95:45.

Gonzalez, Tim, Lucy Hammond, Fred Luntz and Virginia Land. 1999. "Free Love and Nickel Beer: On Throw-away Relationships." *Journal of Sociology and Religion* 12(2):14–29, cited in Emanuel Hiddocke, Cheryl Manson and Ruth Mendez. 2001. *The Death of the American Family.* Upper Saddle River, NJ: Prentice Hall, p. 107.

MAGAZINE ARTICLES. Magazines, which are usually published weekly, bimonthly, or monthly, appeal to the popular audience and generally have a wider circulation than journals. Newsweek and Scientific American are examples of magazines.

Monthly Magazine

Stapleton, Bonnie and Ellis Peters. 1981. "How It Was: On the Trail with Og Mandino." *Lifetime Magazine,* April, pp. 23–24, 57–59.

Weekly or Bimonthly Magazine

Bruck, Connie. 1997. "The World of Business: A Mogul's Farewell." *New Yorker,* October 18, pp. 12–15.

NEWSPAPER ARTICLES

Everett, Susan. 1996. "Beyond the Alamo: How Texans View the Past." *Carrollton Tribune,* February 16, D1, D4.

SOURCES STORED IN ARCHIVES

According to the *ASA Style Guide,* if you refer to a number of archival sources, you should group them in a separate part of the references section and name it "Archival Sources."

Clayton Fox Correspondence, Box 12. July–December 1903. File: Literary Figures 2. Letter to Edith Wharton, dated September 11.

PUBLIC DOCUMENTS

Since the *ASA Style Guide* gives formats for only two types of government publications, the following bibliographical models are based not only on ASA practices but also on formats found in the *CMS* (15.322–411, 16.148–79).

Congressional Journals

References to either the *Senate Journal* or the *House Journal* begin with the journal's title and include the years of the session, the number of the Congress and session, and the month and day of the entry:

> *U.S. Senate Journal.* 1997. 105th Cong., 1st sess., 10 December.

The ordinal numbers *second* and *third* may be represented as *d* (52d, 103d) or as *nd* and *rd*, respectively.

Congressional Debates

> *Congressional Record.* 1930. 71st Cong., 2d sess. Vol. 72, pt. 8.

Congressional Reports and Documents

> U.S. Congress. 1997. House Subcommittee on the Study of Governmental/Public Rapport. *Report on Government Efficiency As Perceived by the Public.* 105th Cong., 2d sess., pp. 11–26.

Bills and Resolutions

Citing a Slip Bill

The abbreviation *S.R.* in the first model below stands for *Senate Resolutions*, and the number following is the bill or resolution number. For references to House bills, the abbreviation is *H.R.* Notice that the second model refers the reader to the more complete first entry. The choice of formats depends upon the one you used in the parenthetical text reference.

> U.S. Senate. 1996. *Visa Formalization Act of 1996.* 105th Cong. 1st sess. S.R. 1437.
>
> *Visa Formalization Act of 1996.* See U.S. Senate. 1996.

Citing the *Congressional Record*

> Senate. 1997. *Visa Formalization Act of 1997.* 105th Cong., 1st sess., S.R. 1437. *Congressional Record* 135, no. 137, daily ed. (10 December): S7341.

Laws

Citing a Slip Law

U.S. Public Law 678. 105th Cong., 1st sess., 4 December 1997. *Library of Congress Book Preservation Act of 1997.*

Library of Congress Book Preservation Act of 1997. U.S. Public Law 678. 105th Cong., 1st sess., 4 December 1997.

Citing the *Statutes at Large*

Statutes at Large. 1998. Vol. 82, p. 466. *Library of Congress Book Preservation Act of 1997.*

Library of Congress Book Preservation Act of 1997. Statutes at Large 82:466.

Citing to the *United States Code*

Library of Congress Book Preservation Act, 1997. U.S. Code. Vol. 38, sec. 1562.

United States Constitution

According to the *CMS* (16.172), the Constitution is not listed in the bibliography.

Executive Department Documents

Department of Labor. 1998. *Report on Urban Growth Potential Projections.* Washington, D.C.: GPO.

The abbreviation for the publisher in the above model, GPO, stands for the Government Printing Office, which prints and distributes most government publications. According to the *CMS* (15.327), you may use any of the following formats to refer to the GPO:

Washington, D.C.: U.S. Government Printing Office, 1984.

Washington, D.C.: Government Printing Office, 1984.

Washington, D.C.: GPO, 1984.

Washington, 1984.

Washington 1984.

Remember to be consistent in using the form you choose.

Legal References

Supreme Court

According to the *CMS* (16.174), federal court decisions are only rarely listed in bibliographies. If you do wish to include such an entry, here is a suitable format:

State of Nevada v. Goldie Warren. 1969. 324 U.S. 123.

For a case prior to 1875, use the following format:

Marbury v. Madison. 1803. 1 Cranch 137.

Lower Courts

> *United States v. Sizemore*. 1950. 183 F. 2d 201 (2d Cir.).

Publications of Government Commissions

> U.S. Securities and Exchange Commission. 1984. *Annual Report of the Securities and Exchange Commission for the Fiscal Year*. Washington, D.C.: GPO.

Publications of State and Local Governments

Remember that references for state and local government publications are modeled on those for corresponding national government documents:

> Oklahoma Legislature. 1991. *Joint Committee on Public Recreation. Final Report to the Legislature, 1990*, Regular Session, on Youth Activities. Oklahoma City.

INTERVIEWS

According to the *CMS* (16.130), interviews need not be included in the bibliography, but if you or your instructor wants to list such entries, here are possible formats:

Published Interview

For an Untitled Interview in a Book

> Jorgenson, Mary. 1998. Interview by Alan McAskill. Pp. 62–86 in *Hospice Pioneers,* edited by Alan McAskill. Richmond, VA: Dynasty Press.

For a Titled Interview in a Periodical

> Simon, John. 1997. "Picking the Patrons Apart: An Interview with John Simon," by Selena Fox. *Media Week,* March 14, pp. 40–54.

Interview on Television

> Snopes, Edward. 1998. Interview by Klint Gordon. *Oklahoma Politicians*. WKY Television, June 4.

Unpublished Interview

> Kennedy, Melissa. 1997. Interview by author. Tape recording. Portland, ME, April 23.

UNPUBLISHED SOURCES

Personal Communications

According to the *CMS* (16.130), references to personal communications may be handled completely in the text of the paper:

> In a letter to the author, dated 16 July, 2001, Mr. Bentley admitted the organizational plan was flawed.

If, however, you wish to include a reference to an unpublished communication in the bibliography, you may do so using one of the following models:

Bentley, Jacob. 1997. Letter to author, 16 July.

Duberstein, Cindy. 2001. Telephone conversation with the author, 5 June.

Timrod, Helen. 1997. E-mail to author, 25 April.

Theses and Dissertations

Hochenauer, Klint. 1980. "Populism and the Free Soil Movement." Ph.D. dissertation, Department of Sociology, Lamont University, Cleveland, OH.

Paper Presented at a Meeting

Zelazny, Kim and Ed Gilmore. 1997. "Art for Art's Sake: Funding the NEA in the Twenty-First Century." Presented at the annual meeting of the Conference of Metropolitan Arts Boards, June 15, San Francisco, CA.

Unpublished Manuscripts

Borges, Rita V. 1993. "Mexican-American Border Conflicts, 1915–1970." Department of History, University of Texas at El Paso, El Paso, TX. Unpublished manuscript.

Working and Discussion Papers

Blaine, Emory and Ralph Cohn. 1995. "Analysis of Social Structure in Closed Urban Environments." Discussion Paper No. 312, Institute for Sociological Research, Deadwood College, Deadwood, SD.

ELECTRONIC SOURCES

CITING ON-LINE SOURCES. The need for a reliable on-line citation system continues to grow, but attempts to establish one are hampered by a number of factors. For one thing, there is no foolproof method of clearly reporting even such basic information as the site's author(s), title, or date of establishment. Occasionally, authors identify themselves clearly; sometimes they place a link to their home page at the bottom of the cite. But it is not always easy to determine exactly who authored a particular cite. Likewise, it can be difficult to determine whether a site has its own title or exists as a subsection of a larger document with its own title. Perhaps the biggest problem facing on-line researchers is the instability of Internet sites. While some cites may remain in place for weeks or months, many either move to another cite—not always leaving a clear path for you to find it—or disappear.

The ASA Style Guide (1997) lists only a few models for electronic sources. You can watch bibliographical history being made on a day-to-day basis on the Internet, where a number of researchers are working to establish viable electronic citation formats. See what you can find, for example, on the following cite on the World Wide Web:

http://www.fis.utoronto.ca/internet/citation.htm

This site offers links to several pages where bibliographers are coming to grips with the problems of Internet referencing. Therefore, until such time as an authoritative citation system for the Internet is available, we suggest the following simple formats, based in part on the models found in the *ASA Style Guide* and in part on the work of other researchers currently available on the Internet.

On-Line Journal Article

The *retrieval date* in the models below is the most recent date on which you accessed the source for your research project.

> Bucknell, Vespasia. 2000. "Servitude as a Way of Life: Religious Denominations in Middle America." *Skeptic's Journal* 4:22–37. Retrieved February 21, 2001 (http://www.religiosk.org/protesta.buck.html).

On-Line Newpaper Article

> Squires, Amanda. 2000. "Hard Times for Social Workers, Says Mayor." *El Paso Sun Times,* July 14, p. 2. Retrieved November 12, 2000 (http://www. elpasosun.com/2000-12/12.html).

The question of whether to break a lengthy site address at the end of a line is not discussed in the *ASA Style Guide,* but one of the guide's models does make such a line break. Other sources suggest breaking a site address only after a slash (/). Do not place a hyphen following the slash. Remember, the one thing that is absolutely required in order to find a site on the Internet is the site address, so make sure that you copy it accurately.

An E-mail Document and Material from Bulletin Boards and Discussion Groups

Due to the ephemeral nature of e-mail sources, most researchers recommend not including citations to e-mail in the bibliography. Instead, you may handle e-mail documentation within the text of the paper.

> In an e-mail dated March 22, 1997, Bennett assured the author that the negotiations would continue.

If, however, you would like to include an e-mail citation in your references section, here is a possible format:

> Bennett, Suzanne. <sbb@mtsu.socla.edu>. 15 March 1997. RE: Progress on education reform petition [e-mail to Courtney Cline <coline@usc.cola.edu>].

The name of the author of the e-mail message is placed first, followed by the author's e-mail address and the date of the message. Next comes a brief statement of the subject of the message, followed by the recipient's name and e-mail address, in brackets.

CITING CD-ROM SOURCE. The publisher of a CD-ROM can usually be identified in the same way as a book's publisher. The following model is for a source with an unascertainable author. Note that it is still necessary to include the latest date on which you accessed the database.

> *Dissertation Abstracts Ondisc.* 1861–1994. CD-ROM: UMI/Dissertation Abstracts Ondisc. Retrieved December 15, 1996.

REFERENCES

Entwisle, Doris, Karl Alexander and Linda Olson. 2000. "Urban Youth, Jobs, and High School." *American Sociological Review* 65(2):279–97.

Johnson, J. D., N. E. Noel and J. Sutter-Hernandez. 2000. "Alcohol and Male Acceptance of Sexual Aggression: The Role of Perceptual Ambiguity." *Journal of Applied Social Psychology* 30(6):1186–1200.

Johnson, William A. and Richard P. Rettig. 1995. "Drug Assessment of Juveniles in Detention." *Social Forces* 28(3):56–69, cited in John Duncan and Mary Ann Hopkins. 2000. "Youth and Drug Involvement: Families at Risk." *British Journal of Addiction* 95:45.

Moore, J. B., Allen Rice and Natasha Traylor. 1998. *Down on the Farm: Culture and Folkways.* Norman, OK: University of Oklahoma Press.

Sczaflarski, Richard. 2001 "The Trumpeter in the Tower: Solidarity and Legend" (in Polish). *World Political Review* 32:79–95.

Squires, Amanda. 2000. "Hard Times for Social Workers, Says Mayor." *El Paso Sun Times,* July 14, p. 2. Retrieved November 12, 2000 (http://www.elpasosun.com/2000-12/12.html).

Stapleton, Bonnie and Ellis Peters. 1981. "How It Was: On the Trail with Og Mandino." *Lifetime Magazine,* April, pp. 23–24, 57–59.

Stomper, Jean. 1973. *Grapes and Rain.* Translated by John Picard. New York: Baldock.

ASA STYLE: SAMPLE REFERENCES PAGE

CITING SOURCES IN APA STYLE

The *Publication Manual of the American Psychological Association* (1994), from which the following formats are taken, suggests that you underline rather than use italics in your paper if it is going to be set in type for publication. The underlined citation material is then italicized when published. However, because most papers constructed from the gidelines of this manual are written as assignments for a particular college class and not for publication, this text will italicize the appropriate citation material.

The APA style uses an *author-date system* of referencing, also known as the *parenthetical-reference system,* which requires two components for each significant reference to a source: (1) a note placed within the text, in parentheses, near where the source material occurs, and (2) a full bibliographical reference for the source, placed in a list of references following the text and keyed to the parenthetical reference within the text. Models for both parenthetical notes (text citations) and full references are given below.

Text Citations in APA Style

Author's Name and Year of Publication in Text

. . . was challenged by Lewissohn in 1999.

Author's Name in Text; Year of Publication in Parentheses

Freedman (1984) postulates that when individuals . . .

Author's Name and Year of Publication in Parentheses

. . . encourage more aggressive play (Perrez, 1979) and contribute . . .

When the citation appears at the end of a sentence, the parenthetical reference is placed inside the period.

. . . and avoid the problem (Keaster, 2000).

Reference Including Page Numbers

Page numbers should only be included when quoting directly from a source or referring to specific passages. Use *p.* or *pp.,* in roman type, to denote page numbers.

Thomas (1961, p. 741) builds on this scenario . . .

. . . in the years to come (Dixon, 1997, pp. 34–35).

Two Authors

When the authors' names are given in the running text, separate them by the word *and*:

> . . . and, according to Holmes and Bacon (1872, pp. 114–116), establish a sense of self.

When the authors' names are given in the parentheses, separate them by an ampersand:

> . . . establish a sense of self (Holmes & Bacon 1872, pp. 114–116).

NOTE: For all sources with more than one author, separate the last two names with the word *and* if they are given within the text and by an ampersand if they appear within parentheses.

Three, Four, or Five Authors

The first citation of a source with three, four, or five authors follows this form:

> . . . found the requirements very restrictive (Mollar, Querley, & McLarry, 1926).

In all subsequent references to a source with three, four, or five authors, place the Latin phrase *et al.*, meaning "and others," after the first author. Note that the phrase appears in roman type, not italics, and is followed by a period.

> . . . proved to be quite difficult (Mollar et al. 1926).

> . . . according to Mollar et al. (1926) . . .

Six or More Authors

In all references to sources with six or more authors, give only the name of the first author, followed by *et al.* (in roman type):

> Kineson et al. (1933) made the following suggestion . . .

When references to two multiple-author sources with the same year shorten to the same abbreviated format, cite as many of the authors as needed to differentiate the sources. Consider these examples:

> Keeler, Allen, Pike, Johnson, and Keaton (1994)

> Keeler, Allen, Schmidt, Wendelson, Crawford, and Blaine (1994)

Using the standard method for abbreviating citations, these two sources would both shorten to the same format:

> Keeler et al. (1994)

However, to avoid confusion, shorten the citations to the above words as follows:

> Keeler, Allen, Pike, et al. (1994)

> Keeler, Allen, Schmidt, et al. (1994)

Group as Author

Use the complete name of a group author in the first citation:

. . . to raise the standard of living (National Association of Food Retailers, 1994).

If the name of the group is lengthy, and if it is easily identified by the general public, you may abbreviate the group name in citations after the first one. In such a case, provide the abbreviation in the first reference, in brackets:

. . . usually kept in cages (Society for the Prevention of Cruelty to Animals [SPCA], 1993).

For subsequent citations:

. . . which, according to the SPCA (1990), . . .

Two Authors with the Same Last Name

. . . the new budget cuts (K. Grady, 1999).

. . . to stimulate economic growth (B. Grady, 2001).

Two Works by the Same Author

If the two citations appear in the same note, place a comma between the publication dates:

George (1992, 1994) argues for . . .

If the two works were published in the same year, differentiate them by adding lower-case letters to the publication dates. Be sure to add the letters to the references in the bibliography, too.

. . . the city government (Estrada, 1994a, 1994b).

Work with No Author Given

Begin the parenthetical reference of a work by an unnamed author with the first few words of the title, either underlining them if they are part of the title of a book, or placing them in quotation marks if they are part of the title of an essay or chapter:

. . . recovery is unlikely (Around the Bend, 2000).

Note that a comma immediately following the underlining is also underlined.

. . . will run again in the next election ("Problems for Smithson," 1996).

Citing Direct Quotations

Direct quotes of fewer than 40 words should be placed in the text, with quotation marks at the beginning and end. The citation should include the page number:

The majority of these ads promote the notion that "If you are slim, you will also be beautiful and sexually desirable" (Rockett & McMinn, 1990, p. 278).

Direct quotes of 40 words or more should be indented five spaces from the left margin and double-spaced. The parenthetical reference following the block quote is placed after the final period:

> During this time there were relative few studies made of the problem, and all of them came to the same sort of conclusion summarized in Brown (1985):
>
> > There are few girls and women of any age or culture raised in white America over the last two generations (or perhaps longer) who do not have at least some significant manifestation of the concerns discussed here, including distortion of body image, a sense of being "out-of-control" in relationship to food, and an addiction to dieting, bingeing, or self-starvation. (p. 61)

The first line of any new paragraph within a block quote should be indented five spaces from the margin of the quote.

Citing Chapters, Tables, Appendixes, Etc.

. . . (Johnson, 1995, chap. 6).

. . . (see Table 4 of Blake, 1985, for complete information).

. . . (see Appendix B of Shelby, 1976).

Citing Reprints

Cite by the original date of publication and the date of the edition you are using:

. . . complaints from Daniels (1922/1976), who takes a different view . . .

Citing More Than One Source in a Reference

Separate citations by a semicolon and place them in alphabetical order by author:

. . . are related (Harmatz, 1987, p. 48; Marble et al., 1986, p. 909; Powers & Erickson, 1986, p. 48; Rackley et al., 1988, p.10; Thompson & Thompson, 1986, p. 62).

Unpublished Materials

If the source is scheduled for publication at a later time, use the designation *in press*:

A study by Barle and Ford (in press) lends support . . .

Personal Communications

Materials such as letters to the author, memos, phone conversations, e-mail, and messages from electronic discussion groups should be cited within the text but not listed among the references. Include in the text note the initials and last name of the person with whom you communicated and give as exact a date as possible:

. . . explained that the work was flawed (P. L. Bingam, personal communication, February 20, 2000).

. . . agrees with the findings and opinions of W. E. Knight (personal communication, October 12, 1998).

Undated Materials

For undated materials, use *n.d.* ("no date") in place of the date:

. . . except that Fox (n.d.) disagrees.

. . . cannot be ascertained (Fox, n.d.).

Classical and Historical Texts

Refer to classical and historical texts, such as the Bible, standard translations of ancient Greek writings, and the *Federalist Papers,* by using the systems by which they are subdivided, rather than the publication information of the edition you are using. Since all editions of such texts employ the standard subdivisions, this reference method has the advantage of allowing your readers to find the cited passage in any published version. You may cite a biblical passage by referring to the particular book, chapter, and verse, all in roman type, with the translation given after the verse number:

"But the path of the just is as the shining light, that shineth more and more unto the perfect day" (Prov. 4:18, King James Version).

The *Federalist Papers* may be cited by their standard numbers:

Madison addresses the problem of factions in a republic (*Federalist* 10).

If you are citing a work whose date is not known or is inapplicable, cite the year of the translation, preceded by the abbreviation *trans.,* in roman type, or the year of the version, followed by the word *version,* in roman:

Plato (trans. 1908) records that . . .

. . . disagrees with the formulation in Aristotle (1892 version).

Newspaper Article with No Author

When citing an unsigned newspaper article, use a shortened form of the title or the entire title if it is short:

. . . painted the new program in glowing colors ("Little Left to Do," 1995).

Public Documents

The APA *Manual* (1994, pp. 224–234) gives detailed information on how to create parenthetical references for public documents. These formats are taken from the 15th edition of *The Bluebook: A Uniform System of Citation* (1991). Here are models for some sources frequently used by psychology professionals.

Legislative Hearings

Information concerning a hearing before a legislative subcommittee is published in an official pamphlet. A parenthetical reference to such a pamphlet begins with a shortened form of the pamphlet's title and includes the year in which the hearing was held:

> . . . city many of the dangers of underfunded school programs (*Funding for Inner City Schools,* 1990).

Bills and Resolutions

Both enacted and unenacted bills and resolutions are cited by their number and house of origin—Senate *(S.)* or House of Representatives *(H.R.),* in roman type—and year. For example, the parenthetical reference to unenacted bill number 7658, originating in the Senate in 1996, would be handled in one of the following ways:

> . . . cannot reject visa requests out of hand (S. 7658, 1996).
>
> . . . according to Senate Bill 7658 (1996).

A parenthetical reference to enacted resolution 94, which originated in the House of Representatives in 1993, is as follows:

> . . . only to U.S. citizens (H.R. Res. 94, 1993).
>
> House Resolution 94 (1993) explains that . . .

Statutes in a Federal Code

In the text, cite the popular or official name of the act and the year:

> . . . in order to obtain a license (Fish and Game Act of 1990).
>
> . . . as provided by the Fish and Game Act of 1990, . . .

Federal Reports

The text and parenthetical references, respectively, to a report from the Senate or House of Representatives are as follows:

> . . . as was finally explained in Senate Report No. 85 (1989), the . . .
>
> . . . was finally clarified (S. Rep. No. 114, 1989).

Court Decisions

> . . . which she failed to meet (*State of Nevada v. Goldie Warren,* 1969).
>
> . . . as was ruled in *State of Nevada v. Goldie Warren* (1969).

Executive Orders

> Executive Order No. 13521 (1993) states that . . .
>
> It was clearly decided (Executive Order No. 13521, 1993) that . . .

References in APA Style

Parenthetical citations in the text point the reader to the fuller source descriptions at the end of the paper known as the references or bibliography. According to the APA *Manual* (1994, p. 174), there is a difference between a reference list and a bibliography of sources consulted for a paper. A reference list gives only those sources used directly in the paper for support, whereas a bibliography may include materials used indirectly, perhaps for background or further reading. Ask your instructor which type of source list you should provide for your class paper.

Like all other parts of the paper, the reference list should be double-spaced. Entries are alphabetized by the first element in each citation. (See the box at the end of the chapter.) The APA reference system uses "sentence-style" capitalization for titles of books and articles, meaning that only the first word of the title and subtitle (if present) and all proper names are capitalized. Titles of periodicals, including journals and newspapers, are given standard, or "headline style," capitalization. In this style all words in a title, except articles *(a, an, the)*, coordinating words *(and, but, or, for, nor)*, and prepositions *(among, by, for, of, to, toward)*, are capitalized.

Although the titles of journals and books are italicized or underlined, titles of chapters or articles are neither italicized, underlined nor enclosed in quotation marks.

NOTE: Title the section *Reference,* in roman type, if there is only one reference in your list. Capitalize only the first letter of the word.

The APA *Manual* (1994, p. 251) requires researchers who are typing a paper that will eventually be published to indent the first line of each item in the reference list five to seven spaces, paragraph style, rather than use the hanging indention common to other bibliographies and reference lists. According to the APA *Manual,* the typesetter will eventually convert the reference entries to the "hanging" style characteristic of most bibliographies, in which all lines of a bibliographical citation after the first line are indented. Ask your instructor which indention style you should use. The examples below use the traditional hanging style—with the first line extended.

BOOKS

One Author

For a single-author source, the author's last name comes first, then the initials of the first and middle names. Add a space after each initial. The date of publication follows in parentheses, followed by a period and then the title of the book, underlined. (Note that the underlining extends past the period.) The city of publication is cited next, then the state (or country if not the United States), unless the city is well-known. Cities that need not be accompanied by state or country include Baltimore, Boston, Chicago, Los Angeles, New York, Philadelphia, San Francisco, Amsterdam, Jerusalem, London, Paris, Rome, Stockholm, and Tokyo. Use postal abbreviations to denote the state (OK, AR, etc.). The name of the publisher is given last:

Northrup, A. K. (1997). *Creative tensions in family units.* Cleveland, OH: Johnstown.

Periods are used to divide most of the elements in the citation, although a colon is used between the place of publication and publisher. Custom dictates that the main title of a book and its subtitle are separated by a colon, even though a colon may not appear in the title as printed on the title page of the book.

Two Authors

Reverse both names, placing a comma after the initials of the first name. Separate the names by an ampersand:

> Spence, M. L., & Ruel, K. M. (1996). *Therapy and the law.* Boston: Tildale.

Three or More Authors

List the names and initials, in reversed order, of all authors of a source:

> Moore, J. B., Macrory, K. L., Rice, A. D., & Traylor, N. P. (1998). *Hanky panky in therapy: Traps for therapists.* Norman, OK: University of Oklahoma Press.

Group as Author

Alphabetize such entries according to the first significant word in the group's name:

> National Association of Physical Therapists. (1994). *Standardization of physical therapy techniques.* Trenton, NJ: Arkway.

Work with No Author Given

Begin the citation with the title of the work, alphabetizing according to the first significant word:

> *Around the bend: Emotional distress among civic administrators.* (1981). Dallas, TX: Turbo.

Editor or Compiler as Author

> Jastow, X. R. (Comp.). (1990). *Saying good-bye: Pathologies in Soviet literature.* New York: Broadus.

> Yarrow, P. T., & Edgarton, S. P. (Eds.). (1987). *Moonlighting in earnest: Second jobs and family instability.* New York: Halley.

Book with Author and Editor

When a book has both an author and an editor, there is no comma between the title and the parentheses enclosing the editor's name, and the editor's last name and initials are not reversed:

> Scarborough, D. L. (1934). *Written on the wind: Psychological maxims* (E. K. Lightstraw, Ed.). Beaufort, SC: Juvenal.

Translated Book

Do not reverse the last name and initials of the translator:

> Zapata, E. M. (1948). *Beneath the wheel: Mental health of the native population in Northern Mexico* (A. M. Muro, Trans.). El Paso, TX: Del Norte.

Untranslated Book

Provide a translation of the title, in brackets, following the title:

> Wharton, E. N. (1916). *Voyages au front* [Visits to the front]. Paris: Plon.

Two Works by the Same Author

Do not use a rule in place of the author's name in the second and subsequent entries; always state the author's name in full and give the earlier reference first:

> George, J. B. (1981). *Who shot John: Psychological profiles of gunshot victims in the Midwest, 1950–1955.* Okarche, OK: Flench & Stratton.
>
> George, J. B. (1989). *They often said so: Repetition and obfuscation in nineteenth-century psychotherapy.* Stroud, OK: Casten.

Author of a Foreword or Introduction

List the entry under the name of the author of the foreword or introduction, not the author of the book:

> Farris, C. J. (1995). Foreword. In B. Givan, *Marital stress among the professariat: A case study* (pp. 1–24). New York: Galapagos.

Selection in a Multiauthor Collection

> Gray, A. N. (1998). Foreign policy and the foreign press. In B. Bonnard & L. F. Guinness (Eds.), *Current psychotherapy issues* (pp. 188–204). New York: Boulanger.

You must provide a complete citation for every selection from a multiauthor collection that appears in the references; do not abbreviate the name of the collection, even if it is included as a separate entry in the reference list.

Signed Article in a Reference Book

> Jenks, S. P. (1983). Fuller, Buckminster. In L. B. Sherman & B. H. Sherman (Eds.), *International dictionary of psychology* (pp. 204–205). Boston: R. R. Hemphill.

Unsigned Article in an Encyclopedia

> Pathologies. (1968). In *Encyclopedia Americana* (Vol. 12, pp. 521–522). Boston: Encyclopedia Americana.

Subsequent Editions

If you are using an edition of a book other than the first, you must cite the number of the edition or the status (such as *Rev. ed.* for revised edition) if there is no edition number:

> Hales, S. A. (1994). *The coming psychological wars* (2d ed.). Pittsburgh, PA: Blue Skies.

> Peters, D. K. (1972). *Social cognition in early childhood* (Rev. ed.). Riverside, CA: Ingot.

Multivolume Work

If you are citing a multivolume work in its entirety, use the following format:

> Graybosch, C. S. (1988). *The rise of psychoanalysis* (Vols. 1–3). New York: Starkfield.

If you are citing only one volume in a multivolume work, use the following format:

> Graybosch, C. S. (1988). *The rise of psychoanalysis: Vol. 1. Bloody beginnings.* New York: Starkfield.

Reprints

> Adams, S. R. (1988). *How to win a promotion: Campaign strategies.* New York: Alexander. (Original work published in 1964)

Modern Editions of Classics

According to the APA *Manual* (1994, p. 173), references to classical texts such as sacred books and Greek verse and drama are usually confined to the text and not given in the references list.

PERIODICALS

JOURNAL ARTICLES. Most journals are paginated so that each issue of a volume continues the numbering of the previous issue. The reason for such pagination is that most journals are bound in libraries as complete volumes of several issues and continuous pagination makes it easier to consult these large compilations.

Journal with Continuous Pagination

While the name of article appears in sentence-style capitalization, the name of the journal is capitalized in standard, or headline, style, and underlined. The underlining includes the volume number, which is separated from the name of the journal by a comma and followed by a comma. Do not use *p.* or *pp.* to introduce the page numbers:

> Hunzecker, J., & Roethke, T. (1987). Teaching the toadies: Cronyism in psychology departments. *Review of Platonic Psychology, 4,* 250–262.

Journal in Which Each Issue Is Paginated Separately

The issue number appears in parentheses immediately following the volume number. In the citation that follows, the quotation marks are necessary only because the title includes a quoted slogan:

> Skylock, B. L. (1991). "Fifty-four forty or fight!": Sloganeering in early America. *American History Digest, 28* (3), 25–34.

English Translation of a Journal Article

If the English translation of a non-English article is cited, give the English title without brackets:

> Sczaflarski, R., & Sczaflarska, K. (1990). The trumpeter in the tower: Solidarity and legend. *World Psychological Review 32,* 79–95.

MAGAZINE ARTICLES. Magazines, which are usually published weekly, bimonthly, or monthly, appeal to the popular audience and generally have a wider circulation than journals. *Newsweek* and *Scientific American* are examples of magazines.

Monthly Magazine

> Stapleton, B., & Peters, E. L. (1981, April). How it was: On the trail with Og Mandino. *Lifetime Magazine, 131,* 24–23, 57–59.

Weekly or Bimonthly Magazine

> Bruck, C. (1997, October 18). The world of private practice: A therapist's farewell. *The New Yorker, 73,* 12–15.

NEWSPAPER ARTICLES. Notice that, unlike in journal or magazine citations, page numbers for references to newspapers are preceded by *p.* or *pp.* (in roman type).

Newspaper Article with No Author Named

> Little left to do before hearing, says new psychology chair. (1996, January 16). *The Vernon Times,* p. A7.

Newspaper Article with Discontinuous Pages

> Everett, S. (1996, February 16). Beyond the Alamo: How Texans view the past. *The Carrollton Tribune,* pp. D1, D4.

PERSONAL COMMUNICATIONS

According to the APA *Manual* (1994, pp. 173–174), personal communications such as letters, memos, and telephone and e-mail messages are cited within the text but do not appear in the reference list.

PUBLIC DOCUMENTS

The APA *Manual* (1994, pp. 224–234) gives detailed information on how to create bibliographical references for public documents. The APA *Manual* cites the 15th edition of *The Bluebook: A Uniform System of Citation* (1991) as the source for this information and refers researchers to this publication for further details. However, some of the sources most commonly used by psychology professionals are presented here.

Legislative Hearings

Information concerning a hearing before a legislative subcommittee is published in an official pamphlet, which is cited as follows:

> *Funding for intelligence testing: Hearing before the Subcommittee on Education Reform of the Education Committee, House of Representatives,* 103d Cong., 2d Sess. 1 (1993).

This citation refers to the official pamphlet reporting on the hearing named, which was held in the U.S. House of Representatives during the second session of the 103d Congress. The report of the hearing begins on page 1 of the pamphlet.

Bills and Resolutions

Bills and resolutions are cited by their number, house of origin—Senate *(S.)* or House of Representatives *(H.R.),* in roman type—and year.

Unenacted Federal Bills and Resolutions

The following citation refers to bill number 1437 from the U.S. Senate, which was created in the first session of the 103d Congress in 1993:

> Visa Formalization Act of 1993, S. 1437, 103d Cong., 1st Sess. (1993).

Enacted Federal Bills and Resolutions

The following citation refers to House Resolution number 192, reported on page 4281 of volume 152 of the *Congressional Record:*

> H.R. Res. 192, 104th Cong., 2d Sess. 152 Cong. Rec. 4281 (1994).

Statutes in a Federal Code

The following entry refers to an act located at section (§) 1043 of title 51 of the *United States Code Annotated,* the unofficial version of the *United States Code:*

Fish and Game Act of 1990, 51 U.S.C.A. § 1043 *et seq.* (West 1993).

The parenthetical reference indicates that the volume of the *United State Code Annotated* in which the statute is found was published in 1993 by West Publishing. The phrase *et seq.,* Latin for "and following," indicates that the act is also mentioned in later sections of the volume.

Federal Reports

The following citation refers to material found on page 4 of the report, which originated in the Senate:

S. Rep. No. 85, 99th Cong., 1st Sess. 4 (1989).

Reports from the House of Representatives begin with the abbreviation *H.R.* instead of *S.*

Court Decisions

Unpublished Cases

The following citation refers to a cased filed in the U.S. Supreme Court on October 3, 1992, under docket number 46-297:

United States v. Vandelay Industries, No. 46-297 (U.S. filed Oct. 3, 1992).

Published Cases

The following citation refers to a case published in volume 102 of the *Federal Supplement,* beginning on page 482:

Jacob v. Warren, 102 F. Supp. 482 (W. D. Nev. 1969).

The decision in the case was rendered by the federal district court for the Western District of Nevada in 1969.

Executive Orders

Executive orders are reported in volume 3 of the *Code of Federal Regulations.* This order appears on page 305:

Exec. Order No. 13521, 3 C.F.R. 305 (1993).

ELECTRONIC SOURCES

CITING ON-LINE SOURCES. The APA is working to establish a standard for citing on-line materials, but the attempt is hampered by a number of factors. For one thing, there is no foolproof method of clearly reporting even such basic information as a Web page's author(s), title, or date of establishment on the Internet. Occasionally, authors identify themselves clearly; sometimes they place a link to their home page at the bottom of the page. But it is not always easy to determine exactly who authored a particular page. Likewise, it can be difficult to determine whether a page has a title of its own or exists as a subsection of another, titled page. Perhaps the biggest problem facing on-line researchers is the instability of Internet sites. While some sites may remain in place for weeks or months, many either move to another location on the Web—not always leaving a clear path for you to find it—or disappear. You can watch bibliographical history being made on a day-to-day basis on the Internet, where a number of researchers are working to establish viable electronic citation formats. See what you can find, for example, on the following site on the World Wide Web:

http://www.fis.utoronto.ca/internet/citation.htm

This site offers links to several pages where bibliographers are coming to grips with the problems of Internet referencing.

The models below, which constitute a step toward establishing a comprehensive and reliable APA referencing system, are based on information from the APA *Manual* (1995, pp. 218–212), which in turn relies on Li and Crane's (1993) *Electronic Style: A Guide to Citing Electronic Information.*

In general, a reference for an on-line source should include as much as possible of the information that would be present in a printed citation, such as the name(s) of the author(s), the date of publication, and title (or titles, if the source is an article within a larger work). If the source is a periodical article, include volume number and page numbers; if the source is a separate publication (such as a book or pamphlet), include the place of publication and publisher. If the source needs further description, provide a brief one, in brackets, as illustrated in the examples below. Complete the citation by noting the date on which you retrieved the source from the Web and the Web address where you found it.

Abstract

Partington, J. W., Sundberg, M. L., Newhourse, L., & Spengler, S. M. (1994). Overcoming an autistic child's failure to acquire a tact repertoire [Abstract]. *Journal of Applied Behavior Analysis, 27*, 733–734. Retrieved June 24, 1999 from the World Wide Web: http://www.apa.org/journals/webref.html

Journal Article

Reid, D. H., & Parsons, M. B. (1995). Comparing choice and questionnaire measures of the acceptability of a staff procedure. *Journal of Applied Behavior Analysis, 28*, 95–96. Retrieved July 7, 1996 from the World Wide Web: http://www.envmed.rochester.edu/wwwrap/behavior/jaba_htm/28/_28-095.htm

E-mail Article

Collarvine, E. T. (1996, June). Consensus thinking: General patterns and idio-syncracies [22 paragraphs]. *Behavioral Sciences Electronic Digest* [On-line serial], *3* (11). Available E-mail: behsci@aix2.ucok.edu Message: Get behsci96-3281

FTP Article

Collarvine, E. T. (1996, June). Consensus thinking: General patterns and idiosyn-cracies [22 paragraphs]. *Behavioral Sciences Electronic Digest* [On-line se-rial], *3* (11). Available FTP: Hostname: bsucok.edu Directory: pub/bsucok/Bsedigest/1996.volume.3 File: Bsedigest.96.3.11.baserate.12.collarvine

For an on-line journal citation, specify the article's length, in brackets, following the title of the article. Also, at the end of the citation specify a document number or ac-cession number to allow for retrieval of the document. Remember, the one thing that is absolutely required in order to find a site on the Internet is the site address, so make sure that you copy it accurately.

NOTE: Because a final period may be misinterpreted as part of the path, do not place one at the end of a citation of an electronic source.

E-mail Messages and Conversations Via Bulletin Boards and Discussion Groups

According to the APA *Manual* (1994, p. 218), these types of materials are cited as per-sonal communications in the text and do not appear in the reference list.

CITING CD-ROM SOURCE

Holly, R. E., & Saffe, I. M. Scent awareness among autistic children: New and contradictory finding [CD-ROM]. *Senses and Sensibility, 3,* 234–251. Ab-stract from: Science IV Archive: Psych Item: 96-0925

References

Adams, D. D., Johnson, T. C., & Cole, S. P. (1989). Physical fitness, body image, and locus of control in first-year college men and women. *Perceptual and Motor Skills, 68,* 400–402.

Becker, H. S. (1963). *The outsiders.* New York: Free Press of Glencoe.

Brown, L. S. (1985). Women, weight, and power: Theoretical and therapeutic issues. *Women and Therapy, 4* (1), 61–71.

Burr, W. R., Leigh, G. K., Day, R. D., & Constantine, J. (1979). Symbolic interaction and the family. In W. R. Burr, F. I. Nye, & I. Reiss (Eds.), *Contemporary theories about the family* (pp. 42–111). New York: The Free Press.

Chavis, C. F. (1997, June). Pumping iron in prison: A hierarchy of values. *Penology: On-Line Issues 2* [On-line journal]. Retrieved November 16, 1997 from the World Wide Web: http://www.penol.usil. edu.htm

Cooley, C. H. (1981). The social self. In T. Parsons, E. Shils, K. D. Naegele, & J. R. Pitts (Eds.), *Theories of society* (pp. 322–338). New York: The Free Press.

Cooley, C. H. (1985). Sex differences in perceptions of desirable body shape. *Journal of Abnormal Psychology, 94* (1), 102–105.

APA STYLE: SAMPLE REFERENCES PAGE

How to Conduct Research in Criminal Justice

Chapter 5   Organizing the Research
Process

Chapter 6   Information Sources and
Distance Learning

5 Organizing the Research Process

GAINING CONTROL

The research paper is where all your skills as an interpreter of details, an organizer of facts and theories, and a writer of clear prose come together. Building logical arguments with facts and hypotheses is the way things get done in criminal justice, and the most successful social scientists are those who master the art of research.

Students new to writing research papers sometimes find themselves intimidated by the job ahead of them. After all, the research paper adds what seems to be an extra set of complexities to the writing process. As any other expository or persuasive paper does, a research paper must present an original thesis using a carefully organized and logical argument. But a research paper often investigates a topic that is outside the writer's own experience. This means that writers of research papers must locate and evaluate information that is new to them, in effect, educating themselves as they explore the topic. A beginning researcher sometimes feels overwhelmed by the basic requirements of the assignment or by the authority of the source material.

In the beginning it may be difficult to establish a sense of control over the different tasks you are undertaking in your research project. You may not know exactly in which direction to search for a thesis, or even where the most helpful sources of information might be located. If you fail to monitor your own work habits carefully, you may unwittingly abdicate responsibility for the paper's argument by borrowing it wholesale from one or more of your sources.

Who Is in Control of Your Paper?

The answer must be *you*—not the instructor who assigned you the paper, and certainly not the published writers whose opinions you solicit. If all your paper does is paste together the opinions of others, it has little use. It is up to you to synthesize an original idea through the evaluation of your source material. Although at the beginning of your research project you will be unsure about many elements of your paper—you will probably not yet have a definitive thesis sentence, for example, or even much understanding of the shape of your argument—you *can* establish a measure of control over the process you will go through to complete the paper. And if you work regularly and systematically, keeping yourself open to new ideas as they present themselves, your sense of control will grow. Here are some suggestions to help you establish and maintain control of your paper.

Understand Your Assignment

A research assignment can fall short simply because the writer did not read the assignment carefully. Considering how much time and effort you are about to put into your project, it is a very good idea to make sure you have a clear understanding of what it is your instructor wants you to do. Be sure to ask your instructor about any aspect of the assignment that is unclear to you, but only after you have thought about it carefully. Recopying the assignment in your own handwriting is a good way to start, even though your instructor may have given the assignment to you in writing. Make sure, before you begin the project, that you have considered the following questions.

What Is Your Topic?

It may be that the assignment gives you a great deal of specific information about your topic, or that you are allowed considerable freedom in establishing one for yourself. In a criminal justice class in which you are studying issues affecting the American criminal justice system, your professor might give you a very specific assignment—a paper, for example, examining the difficulties involved in locating a halfway house in a suburban community—or you may be allowed to choose for yourself the issue that your paper will address. You need to understand the terms, set up in the assignment, by which you will design your project.

What Is Your Purpose?

Whatever the degree of latitude you are given in the matter of your topic, pay close attention to the way in which your instructor has phrased the assignment. Is your primary job to describe a current issue in criminal justice or to take a stand on it? Are you to compare social systems, and if so, to what end? Are you to classify, persuade, survey, analyze? Look for such descriptive terms in the assignment in order to determine the purpose of the project.

Who Is Your Audience?

Your own orientation to the paper is profoundly affected by your conception of the audience for whom you are writing. Granted, your number one reader is your instructor, but who else would be interested in your paper? Are you writing for the citizens of a community? a group of professionals? a city council? A paper that describes the difficulties involved in locating a halfway house in a suburban community may justifiably contain much more technical jargon for an audience of criminal justice professionals than for a citizens group made up of local business and civic leaders.

What Kind of Research Are You Doing?

In your paper you will do one or both of two kinds of research, primary and secondary. *Primary research* requires you to discover information firsthand, often through the conducting of interviews, surveys, or polls. In primary research, you are collecting and sifting through raw data—data not already interpreted by researchers—which you will study, select, arrange, and speculate upon. This raw data may be the opinions of experts or people on the street, historical documents, the theoretical speculations of a famous criminologist, or material collected from other researchers. It is important to carefully set up the method(s) by which you collect your data. Your aim is to gather the most accurate information possible, from which sound observations may be made later, either by you or by other writers using the material you have uncovered.

Secondary research uses published accounts of primary materials. While the primary researcher might poll a community for its opinion on locating a halfway house in their community, the secondary researcher will use the material from the poll to support a particular thesis. Secondary research, in other words, focuses on *interpretations* of raw data. Most of your college papers will be based on your use of secondary sources.

Primary Source	*Secondary Source*
A published collection of Thurgood Marshall's letters	A journal article arguing that the volume of letters illustrates Marshall's attitude toward the media
Material from a questionnaire	A paper basing its thesis on the results of the questionnaire
An interview with the police chief	A character study of the police chief based on the interview

The following is a list of *research approaches* commonly used to study and evaluate crime and other variables in criminal justice.

1. *Comparative research* is used to compare various explanations for crime, delinquency, and other important concerns in criminal justice. It also utilizes cross-cultural analysis. The comparative approach can help generate explanations of how something like crime develops and how society reacts to criminal behavior.
2. *Historical research* evaluates the same society at different times and looks at how a societal component like crime has changed with economic and social development.
3. *Biographical research* employs a case study approach to describe and analyze a certain type of criminal, like a serial killer. The biographical approach can help reveal the needs and motivations of the subject.
4. *Patterns of crime research* help determine where a particular kind of crime is typically committed, who commits it, who is victimized, and what the major dimensions of the criminal act are.
5. *Cohort research* examines the impact of certain cohorts on such subjects as crime and delinquency. The cohort approach is effective in delineating increases and decreases in crime rates. It attempts to isolate changes that are attributable to alterations in attitudes or behavior within an age group.

6. *Records research* uses official and unofficial records to examine such topics as how police arrest suspects, racial discrimination in sentencing, and how parole boards determine the release of inmates.

7. *Survey research* requires firsthand data to be gathered from prepared questions or statements and then often quantified for description or inference. This is usually done in interviews, especially with open-ended questions, or with such direct-sampling techniques as mail-outs or phone surveys. For example, a list of statements about the how police officers should and should not behave might help determine citizens' opinions concerning the role of the police in their community or society. Or in the case of interviews, the researcher might simply ask respondents the open-ended question, "How do you think police officers should behave?"

8. *Experimentation* uses direct observation and measurement to analyze the effects of different treatments on attitudes and behavior. Experiments are designed to control for the influence of outside variables.

9. *Direct observation* of social phenomena is conducted by trained observers who carefully record selected behaviors.

10. *Content analysis* is a method of analyzing written documents that allows researchers to transform nonquantitative data into quantitative data by counting and categorizing certain variables within the data. Content analysts look for certain types of words or references in the texts, then categorize them or count them.

Keep Your Perspective

Whatever type of research you are performing, it is important to keep your results in perspective. There is no way in which you, as a primary researcher, can be completely objective in your findings. It is not possible to design a questionnaire that will net you absolute truth, nor can you be sure that the opinions you gather in interviews reflect the accurate and unchanging opinions of the people you question. Likewise, if you are conducting secondary research, you must remember that the articles and journals you are reading are shaped by the aims of their writers, who are interpreting primary materials for their own ends. The farther you get from a primary source, the greater the possibility for distortion. Your job as a researcher is to be as accurate as possible, and that means keeping in view the limitations of your methods and their ends.

EFFECTIVE RESEARCH METHODS

Establish an Effective Procedure

In any research project there will be moments of confusion, but establishing an effective procedure can prevent confusion from overwhelming you. You need to design a schedule for the project that is as systematic as possible yet flexible enough so that you do not feel trapped by it. A schedule will help keep you from running into dead ends by always showing you what to do next. At the same time, the schedule helps you to retain the presence of mind necessary to spot new ideas and new strategies as you work.

Give Yourself Plenty of Time

There may be reasons why you feel like putting off research: unfamiliarity with the library, the press of other tasks, a deadline that seems comfortably far away. Do not allow such factors to deter you. Research takes time. Working in a library often seems to speed up the clock, so that the hour you expected it to take to find certain sources becomes two hours. You should allow yourself time not only to find material but to read, assimilate, and set it in context with your own thoughts.

The following schedule lists the steps of a research project in the order in which they are generally accomplished. Remember that each step is dependent upon the others, and that it is quite possible to revise earlier decisions in light of later discoveries. After some background reading, for example, your notion of the paper's purpose may change, which may, in turn, alter other steps. One of the strengths of a good schedule is its flexibility. The general schedule lists tasks for both primary and secondary research; you should use only those steps that are relevant to your project.

Task	*Date of Completion*
Determine topic, purpose, and audience	_____
Do background reading in reference books	_____
Narrow your topic; establish a tentative hypothesis	_____
Develop a working bibliography	_____
Write for needed information	_____
Read and evaluate written sources, taking notes	_____
Determine whether to conduct interviews or surveys	_____
Draft a thesis and outline	_____
Write a first draft	_____
Obtain feedback (show draft to instructor, if possible)	_____
Do more research, if necessary	_____
Revise draft	_____
Correct bibliographical format of paper	_____
Prepare final draft	_____
Proofread	_____
Proofread again, looking for characteristic errors	_____
Deadline for final draft	_____

Do Background Reading

Whether you are doing primary or secondary research, you need to know what kinds of work have already been accomplished in your field of study. A good way to start is by consulting general reference works, though you do not want to overdo it (see below). Chapter 6 lists specialized reference works focusing on topics of interest to criminal justice students and professionals. You might find help in such volumes even for specific, local problems, such as how to restructure a juvenile treatment program or plan an antidrug campaign aimed at area schools.

Warning: Be very careful not to rely too exclusively on material taken from general encyclopedias. You may wish to consult one for an overview of a topic with which you are unfamiliar, but students new to research are often tempted to import large sections, if not entire articles, from such volumes, and this practice is not good scholarship. One major reason why your instructor has required a research paper from you is to let you experience the kinds of books and journals in which the discourse of criminal justice is conducted. General reference encyclopedias, such as *Encyclopedia Britanica* or *Collier's Encyclopedia,* are good places for instant introductions to subjects; some encyclopedias even include bibliographies of reference works at the ends of their articles. But you will need much more detailed information about your subject to write a useful paper. Once you have learned what you can from a general encyclopedia, move on.

A primary rule of source hunting is to use your imagination. Determine what topics relevant to your study might be covered in general reference works. If, for example, you are looking for introductory readings to help you with the aforementioned research paper on antidrug campaign planning, you might look into such specialized reference tools as the *Encyclopedia of Social Work* (Edwards 1995). Remember to check articles in such works for lists of references to specialized books and essays.

Narrow Your Topic and Establish a Working Thesis

Before beginning to explore outside sources, it would be a good idea for you to find out what you already know or think about your topic, a job that can only be accomplished well in writing. You might wish to investigate your own attitude toward the topic, your beliefs concerning it, using one or more of the prewriting strategies described in Chapter 1. You might also be surprised by what you know—or don't know—about the topic. This kind of self-questioning can help you discover a profitable direction for your research.

For a research paper in her criminal justice course, Blake Johnson was given the general topic of studying grassroots attempts to legislate morality in American society. She chose the topic of textbook censorship. Here is the course her thinking took as she looked for ways to limit the topic effectively and find a thesis:

General topic:	Textbook censorship
Potential topics:	How a local censorship campaign gets started
	Funding censorship campaigns
	Reasons behind textbook censorship
	Results of censorship campaigns
Working thesis:	It is disconcertingly easy in our part of the state to launch a textbook censorship campaign.

It is unlikely that you will come up with a satisfactory thesis at the beginning of your project. You need a way to guide yourself through the early stages of research toward a main idea that is both useful and manageable. Having in mind a *working thesis*—a preliminary statement of your purpose—can help you select material that is of greatest interest to you as you examine potential sources. The working thesis will probably evolve as your research progresses, and you need to be ready to accept such change. You must not fix on a thesis too early in the research process, or you may miss opportunities to refine it.

Develop a Working Bibliography

As you begin your research, look for published sources—essays, books, and interviews with experts in the field—that may help you with your project. This list of potentially useful sources is your *working bibliography*. There are many ways to discover items for the bibliography. The cataloging system in your library will give you titles, as will specialized published bibliographies in your field. (Some of these bibliographies are listed in Chapter 6.) The general reference works in which you did your background reading may also list such sources, and each specialized book or essay you find will have a bibliography of sources its writer used that may be useful to you.

From your working bibliography you can select items for the final bibliography, which will appear in the final draft of your paper. Early in your research you may not know which sources will help you and which will not. It is important to keep an accurate description of each entry in your working bibliography in order to tell clearly which items you have investigated, which you will need to consult again, and which you will discard. Building the working bibliography also allows you to practice using the required bibliographical format for the final draft. As you list potential sources, include all the information about each source called for by your format, and place the information in the correct order, using the proper punctuation.

The bibliographical format of the American Sociological Association (ASA), a format required for criminal justice papers by many professional journals, is described in detail in Chapter 4 of this manual. The format of the American Psychological Association (APA) is also used in criminal justice journals and is described in Chapter 4.

Write for Needed Information

In the course of your research you may need to consult a source that is not immediately available to you. Working on the antidrug campaign paper, for example, you might find that a packet of potentially useful information is available from a government agency or a public interest group at the state or federal level. Maybe a needed book is not held by your university library or by any other local library. Perhaps a successful antidrug program has been implemented in the school system of a city comparable in size to yours but located in another state. In such situations as these, it may be tempting to disregard potential sources because of the difficulty of consulting them. If you ignore the existence of material important to your project, however, you are not doing your job.

It is vital that you take steps to acquire the needed material. In the first case above, you can simply write the state or federal agency; in the second, you may use your library's interlibrary loan procedure to obtain a copy of the book; in the third, you can track down the council that manages the antidrug campaign by e-mail, mail, or phone and ask for information. Remember that many businesses and government agencies want to share their information with interested citizens; some have employees or entire departments whose job is to facilitate communication with the public. Be as specific as possible when asking for information by mail. It is a good idea to outline your project—in no more than a few sentences—in order to help the respondent determine the types of information that will be useful to you.

Never let the immediate unavailability of a source stop you from trying to consult it. Also, be sure to begin the job of locating and acquiring such long-distance source material as soon as possible, in order to allow for the various types of delays that often occur while conducting a search from a distance.

Evaluate Written Sources

Fewer research experiences are more frustrating than half-remembering something worth using in a source that you can no longer identify. You must establish an efficient method of evaluating the sources listed in your working bibliography. Here are some suggestions:

- By examining a book closely you can usually assess the quality of information presented. The preface and introduction give clues to who the author is, why the work was written, and what methodology and research tools were used in the book's preparation. If the author is an acknowledged authority in the field, this fact will often be mentioned in the preface or the foreword.
- The footnotes, in-text references, and the extent and quality of the bibliography (or in some cases, the lack of one) can also serve as clues about the reliability of the work. If few or no original documents have been used, or if major works in the field have not been cited and evaluated, you have reason to question the quality of the book.
- The reputation of the publisher or organization that sponsors a particular book or periodical says something about its value. Some publishers have rigid standards of scholarship and others do not. For example, the requirements of university presses are generally very high, and the major ones, such as Cambridge, Chicago, Michigan, and Harvard, are discriminating publishers of studies in criminal justice.
- A journal article should announce its intention in its abstract or introduction, which in most cases will be a page or less in length.

This sort of preliminary examination should tell you whether a more intensive examination is worthwhile.

NOTE: Whatever you decide about the source, copy the title page of the book or journal article on a photocopy machine, making sure that all important publication information (including title, date, author, volume number, and page numbers) is included. Write on the photocopied page any necessary information that is not printed there. Without such a record, later on in your research you may forget that you have already consulted a text and find yourself reexamining it.

When you have determined that a potential source is worth closer inspection, explore it carefully. If it is a book, determine whether you should invest the time it will take to read it in its entirety. Whatever the source, make sure you understand not only its overall thesis, but also each part of the argument that the writer sets up to illustrate or prove the thesis. You need to get a feel for the shape of the writer's argument, how the subtopics mesh to form a logical defense of the main point. What do you think of the writer's logic and the examples used? Coming to an accurate appraisal may take more than one reading.

As you read, try to get a feel for the larger argument in which the source takes its place. Its references to the works of other writers will show you where to look for additional material and indicate the general shape of scholarly opinion concerning your subject. If you can see the source you are reading as only one element of an ongoing dialogue instead of an attempt to have the last word on the subject, then you can place the argument of the paper in perspective.

Use Photocopies and Download Articles

Most colleges and universities provide on-line article databases like FIRSTSEARCH and EBSCOHOST, which allow you to download or print out citations or entire articles. Many articles, however, are still not available in these databases, and you may need to photocopy them in the library.

If you do decide to copy source material from on-line or published sources, you should do the following:

- Be sure to follow all copyright laws.
- Have the exact change for the photocopy machines. Do not trust the change machines at the library. They are usually battle-scarred and cantankerous.
- Record all necessary bibliographical information on the photocopy. If you forget to do this, you may find yourself making an extra trip to the library just to get an accurate date of publication or set of page numbers.

Important: Remember that photocopying a source is not the same thing as examining it. You will still have to spend time going over the material, assimilating it in order to use it accurately. It is not enough merely to have the information close at hand or even to read it through once or twice. You should understand it thoroughly. Be sure to give yourself time for this kind of evaluation.

Determine Whether to Conduct Interviews or Surveys

If your project calls for primary research, you may need to interview experts on your topic or to conduct an opinion survey among a select group using a questionnaire. Be sure to prepare yourself as thoroughly as possible for any primary research. Here are some tips:

- Establish a purpose for each interview, bearing in mind the requirements of your working thesis. In what ways might your discussion with the subject benefit your paper? Write down your formulation of the interview's purpose. Estimate the length of time you expect the interview to take and inform your subject. Arrive for your scheduled interview on time and dressed appropriately. Be courteous.
- Learn as much as possible about your topic by researching published sources. Use this research to design your questions. If possible, learn something about the people you interview. This knowledge may help you establish rapport with your subjects and will also help you tailor your questions. Take a list of prepared questions to the interview. However, be ready to depart from your scheduled list of questions in order to follow any potentially useful direction that the interview takes.

- Take notes during the interview. Take along extra pens. The use of a tape recorder may inhibit some interviewees. If you wish to use audiotape, ask for permission from your subject. Follow up your interview with a thank-you letter and, if feasible, a copy of the published paper in which the interview is used.

Draft a Thesis and Outline

Since you will never be able to find and assimilate every source pertaining to your subject, especially if it is a popular or controversial one, you should not prolong your research unduly. You must bring this phase of the project to an end—with the option of resuming it later if the need arises—and begin to shape both the material you have gathered and your thoughts about it into a paper. During the research phase, you have been thinking about your working thesis, testing it against the material you have discovered, considering ways to improve it. Eventually, you must arrive at a formulation of the thesis that sets out an interesting and useful task, one that can be satisfactorily managed within the limits of your assignment and that effectively employs much, if not all, of the source material you have gathered.

Once you have formulated your thesis, it is a good idea to make an outline of the paper. In helping you determine a structure for your writing, the outline is also testing the thesis, prompting you to discover the kinds of work your paper will have to do to complete the task set out by the main idea. Chapter 1 discusses the structural requirements of the formal and the informal outline. (If you have used note cards, you may want to start outlining by first organizing your cards according to the headings you have given them and looking for logical connections among the different groups of cards. Experimenting with structure in this way will lead you to discoveries that further improve your thesis.)

No thesis or outline is written in stone. There is always time to improve the structure or purpose of your paper even after you have begun to write your first draft or, for that matter, your final draft. Some writers actually prefer to do a first draft of the paper before outlining, then study the draft's structure in order to determine what revisions need to be made. *Stay flexible,* always looking for a better connection, a sharper wording of your thesis. The testing of your ideas goes on the entire time you are writing.

Write a First Draft

Despite all the preliminary work you have done on your paper, you may feel resistance to beginning your first draft. Integrating all your material and ideas into a smoothly flowing argument is a complicated task. It may help to think of this first attempt as only a *rough draft,* which can be changed as necessary. Another strategy for reducing reluctance to starting is to begin with the part of the draft that you feel most confident about instead of with the introduction. You may write sections of the draft in any order, piecing the parts together later. But however you decide to start writing—START.

Obtain Feedback

It is not enough that you understand your argument; others have to understand it, too. If your instructor is willing to look at your rough draft, you should take advantage of the opportunity and pay careful attention to any suggestions for improvement. Other readers may be of help, though having a friend or a relative read your draft may not be as helpful as having it read by someone who is knowledgeable in your field. In any event, be sure to evaluate carefully any suggestions you receive for improvement. And always remember, the final responsibility for the paper rests with you.

ETHICAL USE OF SOURCE MATERIAL

You want to use your source material as effectively as possible. This will sometimes mean that you should quote from a source directly, while at other times you will want to express source information into your own words. At all times, you should work to integrate the source material skillfully into the flow of your written argument.

When to Quote

You should quote directly from a source when the original language is distinctive enough to enhance your argument, or when rewording the passage would lessen its impact. In the interest of fairness, you should also quote a passage to which your paper will take exception. Rarely, however, should you quote a source at great length (longer than two or three paragraphs). Nor should your paper, or any substantial section of it, be merely a string of quoted passages. The more language you take from the writings of others, the more disruptive the quotations are to the rhetorical flow of your own language. Too much quoting creates a choppy patchwork of varying styles and borrowed purposes in which your sense of your own control over the material is lost.

ACKNOWLEDGE QUOTATIONS CAREFULLY

Failing to signal the presence of a quotation skillfully can lead to confusion or choppiness:

> The U. S. Secretary of Labor believes that worker retraining programs have failed because of a lack of trust within the American business culture. "The American business community does not visualize the need to invest in its workers" (Winn 1992:11).

The first sentence in the above passage seems to suggest that the quote that follows comes from the Secretary of Labor. Note how this revision clarifies the attribution:

> According to reporter Fred Winn (1992), the U. S. Secretary of Labor believes that worker retraining programs have failed because of a lack of trust within the American business culture. Summarizing the Secretary's view, Winn writes, "The American business community does not visualize the need to invest in its workers" (p. 11).

The origin of each quote must be signaled within your text at the point where the quote occurs, as well as in the list of works cited, which follows the text. Chapter 4 describes documentation formats set forth by the American Sociological Association (ASA) and the American Psychological Association.

QUOTE ACCURATELY

If your quotation introduces careless variants of any kind, you are misrepresenting your source. Proofread your quotations very carefully, paying close attention to such surface features as spelling, capitalization, italics, and the use of numerals. Occasionally, in order either to make a quotation fit smoothly into a passage, to clarify a reference, or to delete unnecessary material from a quotation, you may need to change the original wording slightly. You must signal any such change to your reader by using brackets:

> "Several times in the course of his speech, the attorney general said that his stand [on gun control] remains unchanged" (McAffrey 1995:2).

Ellipses may be used to indicate that words have been left out of a quote:

> "The last time voters refused to endorse one of the senator's policies . . . was back in 1982" (Law 1992:143). This probably means that . . .

When you integrate quoted material with your own prose, it is unnecessary to begin the quote with ellipses:

> Benton raised eyebrows with his claim that "nobody in the mayor's office knows how to tie a shoe, let alone balance a budget" (Williams 1990:12).

Paraphrasing

Your writing has its own rhetorical attributes, its own rhythms and structural coherence. Inserting too many quotations into a section of your paper can disrupt the patterns you establish in your prose and diminish the effectiveness of your own language. Paraphrasing, or recasting source material in your own words, is one way of avoiding the risk of creating a choppy hodgepodge of quotations. Paraphrasing allows you to communicate ideas and facts from a source in your own prose, thereby keeping intact the rhetorical characteristics that distinguish your writing.

Remember that a paraphrase is to be written in *your* language; it is not a near copy of the source writer's language. Merely changing a few words of the original does justice to no one's prose and frequently produces stilted passages. This sort of borrowing is actually a form of plagiarism. In order to fully integrate the material you wish to use into your writing, *use your own language*.

Paraphrasing may actually increase your comprehension of source material because in recasting a passage you have to think very careful about its meaning, more carefully, perhaps, than you might if you merely copied it word for word.

Avoiding Plagiarism

Paraphrases require the same sort of documentation as direct quotes. The words of a paraphrase may be yours, but the idea belongs to someone else. Failure to give that person credit, in the form of references within the text and in the bibliography, may make you vulnerable to a charge of plagiarism.

Plagiarism is the using of someone else's words or ideas without proper credit. While some plagiarism is deliberate, produced by writers who understand that they are guilty of a kind of academic thievery, much of it is unconscious, committed by writers who are not aware of the varieties of plagiarism or who are careless in recording their borrowings from sources. Plagiarism includes:

- Quoting directly without acknowledging the source
- Paraphrasing without acknowledging the source
- Constructing a paraphrase that closely resembles the original in language and syntax.

One way to guard against plagiarism is to keep careful records in your notes of when you have actually quoted source material directly and when you have paraphrased—making sure that the wording of the paraphrase is yours. Make sure that all direct quotes in your final draft are properly set off from your own prose, either with quotation marks or in indented blocks.

What kind of paraphrased material must be acknowledged? Basic material that you find in several sources need not be acknowledged by a reference. For example, it is unnecessary to cite a source for the information that Franklin Delano Roosevelt was elected to a fourth term as President of the United States shortly before his death, because this is a commonly known fact. However, Professor Smith's opinion, published in a recent article, that Roosevelt's winning of a fourth term hastened his death is not a fact but a theory based on Smith's research and defended by her. If you wish to use Smith's opinion in a paraphrase, you need to credit her, as you should all judgments and claims from another source. Any information that is not widely known, whether factual or open to dispute, should be documented. This includes statistics, graphs, tables, and charts taken from a source other than your own primary research.

Information Sources and Distance Learning

LIBRARY RESEARCH

Unless your writing assignment involves studying a particular institution or program, you will probably find most of the information you need for your criminal justice papers in your college library or on the internet. This chapter provides you with some information that will help you get started in both these places, and a guide to prospective and current distance learners.

A good place to start your library research is with one of the many reference books you will find there. There are three principal categories of reference works:

1. *Finding aids* help you locate publications that contain information on your topic. Finding aids include bibliographies and periodical indexes.
2. *Content reference works* contain the type of factual information you are looking for about a particular topic or topics. Content reference works include handbooks, yearbooks, subject dictionaries, and subject encyclopedias.
3. *Guides to the literature* are books or articles that list—and usually describe—reference works that fit in one or both of these two categories. Some guides include discussions of various types of research materials, such as government publications, while others include lists of important book-length studies on topics in a subject field. Researchers can identify criminal justice reference publications by consulting a guide that covers a wide spectrum of fields related to their areas of interest.

A good example of a general reference guide is Bray's *Guide to Reference Books* (1996). Bray and his associates have compiled one of the most complete guides to reference sources in criminal justice and other areas.

A handbook is a compact factbook designed for quick reference. Handbooks, sourcebooks, and factbooks usually deal with one broad subject area and emphasize generally accepted data rather than recent findings. In the latter respect, handbooks differ from yearbooks, although these reference tools overlap in the way they are used and the information they include. Two types of handbooks useful to criminal justice students are: (1) statistical handbooks, which provide data about a number of demographic and social characteristics, and (2) subject handbooks, which offer a comprehensive summary of research findings and theoretical propositions for broad substantive areas in a discipline. Among the many statistical handbooks useful to students and professionals in criminal justice is the U.S. Bureau of the Census; *Historical Statistics of the United States: Colonial Times to 1970* (1971). This two-volume work contains statistics on a wide spectrum of social and economic developments from the colonial period to 1970. An example of a subject handbook is Ronald F. Becker's *Specific Evidence and Expert Testimony Handbook: A Guide for Lawyers, Criminal Investigators and Forensic Specialists* (1997).

Yearbooks contain a good deal of background information, but they are primarily factbooks that focus on the developments and events of a given year. Unlike handbooks, they emphasize current information. Like handbooks, there are two types of yearbooks most useful for criminal justice students: (1) statistical yearbooks, which provide the most recent data on social and demographic characteristics, and (2) subject yearbooks, which review current theory and research.

Indexes contain lists of citations of articles printed in journals, magazines, and other periodicals. The standard citation for articles in indexes includes the author's name, the date of the issue in which the article appears, the article title, the name of the journal, the volume and/or issue number, and page numbers. Abstracts contain short summaries of articles or books. Indexes and abstracts are important because the articles in scholarly journals often update information found in books or in some cases constitute the only published treatments of certain topics.

Specialized indexes and abstracts list articles published in scholarly journals by subject and author. Most are now retrievable through computer on-line database systems located in the library. These CD-ROM networks index thousands of professional and popular articles in most academic areas. The most useful databases for criminal justice are stored in *Sociofile* and *Social Sciences Index*. *Sociofile* indexes over 1900 international journals in criminal justice and related fields that are stored in two print indexes titled *Sociology Abstracts* (1974–present) and *Social Planning, Policy and Development Abstracts*. *Social Sciences Index* (1983–present) indexes journals in most of the social sciences, including anthropology, area studies, criminal justice, economics, environmental science, geography, law, political science, and sociology.

Other databases that offer potential sources for criminal justice studies are:

- *ERIC,* an education database (1966–present) consisting of the *Resources in Education* file and the *Current Index to Journals in Education* file, compiling journal article citations with abstracts from over 750 professional journals
- *Psyclit,* a psychology database (1974–present) that compiles summaries of literature in psychology and related disciplines and corresponds to *Psychological Abstracts,* which indexes about 1300 professional journals in 27 languages
- *Reader's Guide to Periodical Literature,* a popular and general interest database (1983–present) that provides citations from more than 900 journals and magazines in the popular press
- *United States Government Periodical Index,* a government periodicals database (November 1994–present) that indexes approximately 180 U.S. government sources

covering a wide variety of subjects. The reference department in many libraries also allows you to access over 40 other databases on FIRSTSEARCH.

Although indexes in general identify journal articles according to broad topical areas without a discipline focus, discipline indexes identify articles in journals by discipline or group of related disciplines (such as social sciences) instead of by topic. An example of a discipline index that indexes by subject articles in social science, economics, and anthropology journals as well as those in criminal justice and sociology is the *Social Sciences Index* (annual). A quarterly publication with annual cumulations, this index organizes—by subject and author—articles in some 260 journals in anthropology, sociology, law, and criminology. A helpful feature is the separate "Book Reviews" index at the back of each issue.

Academic books, along with articles from professional journals, will usually form the greater part of the criminal justice student's reading list on an individual research topic. If the sources of information used in book research are unreliable, the results will be unsatisfactory. There are two principal paths for a student to take in evaluating a book-length study: (1) rely upon book reviews, and (2) examine the bibliographic character of the book itself. Most professional journals contain a book review section where scholars in the field present critiques of books recently published. These reviews usually give an accurate assessment of the book's quality from a criminal justice standpoint. When faced with a choice of several books, you can save time by reading book reviews to select the most useful, authoritative sources.

Since 1975, reviews appearing in most of the major journals that are germane to criminal justice have been indexed in the book review section of the *Social Sciences Index*. The reviews are indexed by the name of the author of the book, the journal, volume, date, and the page(s) of the review. There is a time lag, however, sometimes more than a year, between a book's publication and the appearance of a review in a scholarly journal.

U.S. government publications comprise all the printed public documents of the federal government. The materials include, for example, the official records of Congress; the text of laws, court decisions, and public hearings; rulings of administrative and regulatory agencies; studies of economic and social issues commissioned by official agencies; and the compilation of statistics on a number of social and demographic characteristics.

Federal publications provide students and professionals with material for research in many criminal justice subfields, such as the educational attainment of minorities, sex discrimination, and drug and physical abuse. But *beware:* The fact that a document is "official" is no automatic guarantee of the accuracy of the information or data it contains. Accuracy depends on the methods of information and data collection the agency uses. Therefore, the use of these publications, like the use of any other source material, requires good judgment. You can identify relevant late nineteenth- and twentieth-century federal government publications by consulting:

U. S. Superintendent of Documents. Annual. *Monthly Catalog of United States Government Publications*. Washington, D.C.: Government Printing Office.

CRIMINAL JUSTICE ON THE WORLD WIDE WEB

The number of criminal justice sources on the Internet is bewildering. This chapter provides a few good places to begin your search.

The home page of your college's criminal justice department is often a good place to start your search. The Department of Criminal Justice of the State University of New York at Albany, for example, features a criminal justice resource page <http://www.albany.edu/scj/links.html>. Each of its topical links bring you to sites that give you just about everything you might want to write a typical term paper:

National and State Laws

Court Sites

Federal Government

State Government

National Organizations

International Government

International Organizations and Information

Policing and Crime Prevention

Journals

Discussion Lists

Libraries and Information Providers

Statistics and Data Providers

State Correctional Institutions

Prison Information

Death Penalty Information

Drugs in America

Criminal Justice Education

Victims of Crime

Sex Offender Registries

Restorative Justice

More Guides to Criminal Justice on the Internet

If you select, for example, the link entitled *federal government,* you will find a link to the United States Department of Justice. The Department of Justice (DOJ) home page <http://www.usdoj.gov/> lists the following links, each of which features a broad array of information:

American Indian and Alaska Native Affairs Desk (OJP)

Antitrust Division

Attorney General

Bureau of Justice Assistance (OJP)

Bureau of Justice Statistics (OJP)

Civil Division

Civil Rights Division

Community Oriented Policing Services (COPS)

Community Relations Service

Corrections Program Office (OJP)

Criminal Division

Diversion Control Program (DEA)

Drug Courts Program Office (OJP)

Drug Enforcement Administration

Environment and Natural Resources Division

Executive Office for Immigration Review

Executive Office for U.S. Attorneys

Executive Office for U.S. Trustees

Executive Office for Weed and Seed (OJP)

Federal Bureau of Investigation

Federal Bureau of Prisons

Foreign Claims Settlement Commission of the United States

Immigration and Naturalization Service

INTERPOL—U.S. National Central Bureau

Justice Management Division

National Criminal Justice Reference Service (OJP)

National Drug Intelligence Center

National Institute of Corrections (FBOP)

National Institute of Justice (OJP)

Office of the Associate Attorney General

Office of the Attorney General

Office of Attorney Personnel Management

Office of Community Dispute Resolution

Office of the Deputy Attorney General

Office of Dispute Resolution

Office of Information and Privacy

Office of the Inspector General

Office of Intelligence Policy and Review

Office of Justice Programs

Office of Juvenile Justice and Delinquency Prevention (OJP)

Office of Legal Counsel

Office of Legislative Affairs

Office of the Pardon Attorney

Office of Policy Development

Office of Professional Responsibility

Office of Public Affairs

Office of the Solicitor General

Office for State and Local Domestic Preparedness Support (OJP)

Office of Tribal Justice

Office for Victims of Crime (OJP)

Tax Division

U.S. Attorneys

U.S. Marshals Service

U.S. Parole Commission

U.S. Trustee Program

Violence Against Women Office (OJP)

If you select the link titled Bureau of Justice Statistics, <http://www.ojp.usdoj.gov/bjs/> you will find statistics about virtually every aspect of criminal justice:

Crimes and victims—Criminal victimization, general, Characteristics of crimes, Characteristics of victims, . . .

Drugs and crime

Homicide trends

Criminal offenders

Special topics—Sourcebook of criminal justice statistics, Firearms and crime, World Factbook of Criminal Justice Systems, . . .

Crime and Justice data from other sources—FBI's Uniform Crime Reports, juvenile justice statistics, international statistics, . . .

The Justice System

Law Enforcement—Federal, State and local, . . .

Prosecution

Courts and Sentencing—Pretrial release and detention, Criminal case processing, Criminal sentencing, Federal justice, Civil justice, Court organization, Indigent defense

Corrections—Capital punishment, Jails, Prisons, Probation and parole

Expenditure and Employment

Once you have begun, you will discover thousands of sites and dozens of types of information that we have not mentioned in this chapter. Some of the things you will find will include:

• *Mailing Lists.* You can join a mailing list and receive by e-mail or in printed form publications from a wide variety of organizations.

- *Bibliographies.* Numerous extensive bibliographies of criminal justice information already appear on the net.
- *Publishers and Bookstores.* You will find publishers offering to sell you virtually any title in criminal justice, and you will find bookstores that offer not only new books, but old, outdated, and rare editions of many texts.
- *Criminal Justice Projects.* Many criminal justice research and discussion projects have home pages on the net, and sometimes you can join in the research and discussions.
- *Criminal Justice Resource Guides.* Guides to all these resources and more are on the net, with clickable links which take you directly to the sources they cite.
- *News Groups.* News groups are Internet pages in which people exchange information on current events.

A GUIDE TO DISTANCE LEARNING

For Students Considering Distance Learning

Perhaps you are apprehensive about taking a distance learning course, or you want to take one but simply do not know where to begin. In either case, this introduction will help you. You will have some important questions to ask before you sign up, and this section will address some of them.

Are distance learning courses effective? Initial studies indicate that if the amount of material learned is a valid criterion for effectiveness, then the answer is yes. After reviewing more than 400 studies of the effectiveness of distance learning courses, Thomas L. Russell, director emeritus of instructional telecommunications at North Carolina State University, concluded that distance learning and classroom courses were equally effective (Young 2000).

This does not mean, however, that the two methods are the same in every respect. Texas Tech psychology professors Ruth S. and William S. Maki found that when they compared distance learning and classroom introductory psychology courses, distance learning students scored from 5 to 10 percent higher on tests of knowledge, but they expressed less satisfaction with their courses (Carr 2000). Students in classroom courses appreciated more contact with their professors, and their distance learning counterparts observed that on-line courses required more work than their comparable classroom experiences.

Have no fear. If you are a bit uneasy about taking a course in which your only contact with people will be through e-mail or over the Internet, you have a lot of company. New experiences are almost always a bit unsettling, and you may not be as comfortable with a computer as some of your friends. The good news is that institutions that provide distance learning have gone to a lot of trouble to make your introduction to their courses as trouble-free as possible. Entering their websites you will find easy-to-follow, step-by-step instructions and sources of help on every aspect of your new education experience.

You may want to visit such a site to see what it is like. A good example is World Campus 101 (WC101), established by Penn State University and located at the following address: <www.worldcampus.psu.edu>. (Note that in this chapter, as throughout this manual, web addresses that appear within the text are enclosed in angle brackets [< >], which are not part of the address.) WC101 is itself a short course that introduces you to Penn State's on-line courses. Presented in five "modules," WC101 covers the following topics, which include virtually everything you would need to know to take an on-line course:

1. Learning how to be a Penn State World Campus student
2. Using on-line course materials
3. Interacting with our instructor and fellow students
4. Using academic resources in your courses
5. Getting help when you need it.

Another good site to visit is <www.onlinelearning.net>, a service of UCLA Extension. This site features answers to a list of commonly asked questions. Reading this material will help you overcome some of your initial trepidation. With respect to UCLA's on-line courses, you will find, for example, that UCLA extension on-line courses:

- are open to anyone
- have specific dates of course initiation and completion
- require printed textbooks
- can be taken anywhere that a fairly recent personal computer can operate
- are accompanied by technical support to help when problems arise
- are given "asynchronously," which is to say that students are required to send messages frequently to the instructor and other students and can do so at any time of the day or night
- have actual instructors who do all the same things over the website that instructors in classroom courses do
- have enrollments that are normally limited to twenty students, allowing adequate access to the instructor
- feature special software that is provided by UCLA Extension, including an on-line orientation to this software.

Are you likely to succeed in distance learning courses? The answer to this question depends on a number of factors, and every student will react to distance learning situations at least a little differently than any other. Among the factors that will influence your chances of success, however, are how comfortable and happy you are about:

- working alone
- communicating with people without seeing them
- accomplishing tasks without reminders from others
- using computers
- solving occasional technical problems on computers
- learning how to use new software.

How is distance learning different from classroom courses? Everything considered, distance learning and classroom courses are probably more alike than they are different. Like classroom courses, distance learning courses have an actual person as an instructor, actual people as students, and printed or printable course materials. In both classroom and distance learning, individual initiative and responsibility are required for success, and in both settings the quality of the course depends, in large part, on the competence of the instructor.

The primary differences are that in distance courses you will work alone on a computer, and you will spend your course-related time according to your own schedule rather than attending classes. While in classroom courses other students and the instructor have a physical presence, in distance learning your contact with others is in electronic form. Interestingly, many students report spending more time on their distance learning courses than on their classroom courses.

Distance learning, therefore, offers several advantages over regular classroom courses. You don't need to commute or relocate. Your distance learning schedule can vary from day to day and week to week. You can connect on a whim or wait until something awakens you at 2 A.M. and you are unable to get back to sleep. In addition, the interaction with other students in on-line courses is often more satisfying than you might first suspect. As messages start streaming back and forth, each student's personality is revealed. Some students send photos of themselves so that you have a better idea of who they are.

There is also a down side to distance learning. One disadvantage to on-line learning which seems to irritate distance learners the most is that they cannot get instant feedback; you can't just raise your hand and receive an immediate answer to your questions, as you can in a classroom. A related drawback—subtle but profound—is that the nonverbal responses that students unwittingly come to count on in a classroom are missing from an on-line course. Is your on-line instructor frowning or smiling as she makes a certain comment? In other words, the act of communication is sometimes more complex than we think. Sometimes on-line course instructions are not sufficiently focused or specific, and it may take several communications to understand an assignment.

Another potential difficulty with distance learning is that on-line students are less likely to appreciate options than students in classrooms. Rather than welcome the chance to make their own choices, they tend to want to do exactly what the instructor wants.

Other problems occasionally appear in on-line courses. Sometimes course materials provide ambiguous instructions and out-of-date hyperlinks. Testing can be complicated and may require special passwords. Some students must go to their local community college to take examinations, but other on-line colleges simply remind students of their academic integrity statements. If you are social by nature you may suffer from feelings of isolation. You may find that it takes longer to establish rapport with on-line students with whom you have little in common. There may be some initial confusion as you learn how to run the system and interact effectively, or you may have difficulty interpreting messages from other students. As with your on-line teacher, so with your classmates: Lack of visual contact means a loss of inflection. Humor and sarcasm are more difficult to detect in written communications. And finally, you may face what seems at times to be an overwhelming volume of e-mail featuring a lot of repetition (Hara and King 1999).

This brief survey of characteristics of distance learning may help you deal with a range of on-line situations as they arise. All in all, if you assess your own personality correctly, your chances of success in distance learning are substantial.

Criminal Justice Distance Learning Courses and Distance Learning Resources On-line

New distance learning courses are appearing daily. In its second survey of distance education programs, the U.S. Department of Education (1998) identified 1680 programs, offering 54,000 on-line courses and enrolling 1.6 million students. These figures represent a 72 percent increase in distance learning activity from 1995 to 1998. The number of criminal justice courses offered on-line, however, is not extensive. Here are some tips for locating distance learning courses in criminal justice.

Your local bookstore (as well as *amazon.com* and *barnesandnoble.com*) will offer several guides to distance learning. Among those currently available are:

- Bear, John and Mariah Bear. 2001. *Bear's Guide to Computer Degrees by Distance Learning.* New York: Ten Speed Press.
- Kramer, Candice. 2001. *Success in Distance Learning.* New York: Delmar Publishers.
- Lockwood, Fred and Anne Gooley. 2001. *Innovations in Open and Distance Learning: Successful Development of Online and Web-Based Learning.* New York: Stylus Publishing.
- Picciano, Anthony G. 2001. *Distance learning: Making Connections across Virtual Space and Time.* Upper Saddle River, NJ: Prentice Hall.

Your Internet search for a suitable course may take some time, since offerings change continuously. You can find links to many colleges and universities at *Web U.S. Universities, by State* (<www.utexas.edu/world/univ/state/>). You will also find that Western Governors University (<www.wgu.edu>) provides a list of criminal justice courses available at several other colleges. The list of general distance education resources on the Internet changes almost daily, but some you may want to examine are:

American Distance Education Consortium	www.adec.edu
Chronicle of Higher Education	www.chronicle.com
Distance Education at a Glance	www.uidaho.edu/evo/distglan.html
Distance Education Clearing House	www.uwex.edu/disted/home.html
Distance-Educator.com	www.distance-educator.com
International Center for Distance Learning	www.icdl.open.ac.uk
Web Based Learning Resources Library	www.outreach.utk.edu/weblearning/
World Lecture Hall	www.utexas.edu/world/lecture/

For Students About to Take or Taking On-Line Courses

Since you have decided to take a distance learning course, decide to study effectively. Studying for distance learning courses requires the same sort of discipline as studying for classroom courses, with one notable difference. For some people, class attendance is energizing. It helps stimulate their desire to study. This stimulus is, of course, absent for distance learners, but e-mail communication with other students and the instructor may serve the same purpose for some. In general, the same study habits that lead to success in regular courses also lead to success in on-line courses.

Morgan (1991) has identified two approaches that students take to distance learning. In the first less effective method, which Morgan calls the *surface* approach, students tend to focus on the signs. This is to say that they see the trees rather than the forest. They concentrate on the text or instruction itself rather than on catching the idea or spirit of what is going on. They get stuck on specific elements of the task rather than understanding the whole task. Less effective students like to memorize data, rules, and procedures, which becomes a crutch and substitutes for the more important task of understanding concepts. They also associate concepts and facts in an unreflective manner, failing to understand how specific facts are related to certain concepts, and therefore getting principles confused with evidence for those principles. Moreover, they consider assignments as mere tasks, or requirements imposed by the instructor, instead of seeing assignments as ways to learn skills or understand concepts that meaningfully relate to the goals of the course or to the realities of life.

As an alternative to this surface approach, Morgan proposes a *deep* approach, in which the student focuses on the concepts being studied and on the instructor's arguments as opposed to the tasks or directions for assignments. The deep approach encourages students to relate new ideas to the real world, to constantly distinguish evidence (data) from argument (interpretations of data), and to organize the course material in a way that is personally meaningful.

Brundage, Keane, and Mackneson (1993) have found that successful distance learners are able to:

- Assume responsibility for motivating themselves
- Maintain their own self-esteem irrespective of emotional support that may or may not be gained from the instructor, other students, family, or friends
- Understand accurately their own strengths and limitations and become able to ask for help in areas of weakness
- Take the time to work hard at effectively relating to the other students
- Continually clarify for themselves and others precisely what it is that they are learning and become confident in the quality of their own observations
- Constantly relate the course content to their own personal experience.

One final thought: Studies indicate that the dropout rate for distance learners is higher than for traditional courses. In part this is because distance learners tend to underestimate their other obligations and the time it will take to successfully complete their on-line course. Before you begin, be sure that you allow enough time to not only complete your course, but to do so with a reasonable measure of enjoyment. Good luck in your adventure in distance education. Armed with the information in this introduction, your chances of success are good.

Writing Different Types of Criminal Justice Papers

Chapter 7   Brief Writing Assignments

Chapter 8   Criminal Justice Agency
                  Case Studies

Chapter 9   Criminal Justice Policy
                  Analysis Papers

Brief Writing Assignments

REACTION PAPERS

The purpose of this assignment is to develop and sharpen your critical thinking and writing skills. Your objective in writing this assignment is to define an issue clearly and to formulate and clarify your position on that issue by reacting to a controversial statement.

Completing this assignment requires accomplishing the following six tasks:

1. Select a suitable reaction statement.
2. Explain your selection.
3. Clearly define the issue addressed in the statement.
4. Clearly state your position on the issue.
5. Defend your position.
6. Conclude concisely.

Select a Suitable Reaction Statement

Your first task is to find or write a statement to which to react. Reaction statements are provocative declarations. They are controversial assertions that beg for either a negative or a positive response. Your instructor may assign a reaction statement, you may find one in a newspaper or on the Internet or hear one on television, or you may construct one yourself, depending on your instructor's directions. The following statements may elicit a polite reply but will probably not stir up people's emotions. They are, therefore, not good reaction statements:

- It's cold out today.
- Plants are beautiful.
- Orange is not green.
- Saturday morning is the best time to watch cartoons.

The following statements, however, have the potential to be good reaction statements because when you hear them you will probably have a distinct opinion about them:

- Abortion is murder.

- Capital punishment is necessary.
- Government is too big.

Such statements are likely to provoke a reaction, either negative or positive, depending on the person who is reacting to them. While they may be incendiary, they are also both ordinary and vague. If your instructor assigns you a statement to which to react, you may proceed to the next step. If you are to select your own, select or formulate one that is provocative, imaginative, and appropriate to the course for which you are writing the paper. Professor Rettig, for example, once assigned this statement in his Innovations in Corrections class:

Prisons should be run by the private sector of society.

Consider the following examples of reaction statements for other criminal justice classes:

- Automatic weapons should be banned.
- Criminals should be punished, not rehabilitated.
- Juveniles who commit heinous crimes should be certified and tried as adults.
- Criminal justice agencies ought to be run more like businesses.
- The Supreme Court should take an active role in determining social policy.
- Police officers should not carry weapons of deadly force.

Where do you find good reaction statements? A good way is to think about subjects that interest you. When you hear something in class that sparks a reaction because you either agree or disagree with it, you know you are on the right track. Be sure to write your statement and ask your instructor for comments on it before beginning your paper. Once you have completed your selection, state it clearly at the beginning of your paper.

Explain Your Selection

After you have written the reaction statement, write a paragraph that explains why it is important to you. Be as specific as possible. Writing "I like it" does not tell the reader anything useful, but sentences like the following are informative: "Innocent people are being shot down by violent gangs in the inner city. We must crack down on gang violence in order to make the inner city safe for all who live there."

Clearly Define the Issue Addressed in the Statement

Consider the statement assigned by Professor Rettig: " Prisons should be run by the private sector of society." What is the most important issue addressed in this statement? Is it the notion that punishment and/or rehabilitation can be handled more efficiently by private business? Or is it the question of whether or not government can run a prison as economically as private business? Perhaps some aspects of the statement are more important. As you define the issue addressed in the statement, you provide yourself with some clarification of the statement that will help you state your position.

Clearly State Your Position on the Issue

In response to Professor Rettig's statement, you might begin by saying, "Much of the waste and corruption could be omitted if prisons were run like private businesses. The money saved from this transformation in prison management could be returned to taxpayers in the form of tax cuts." The reader of this response will have no doubt about where you stand on this issue.

Defend Your Position

You should make and support several arguments to support your stand on the issue. When evaluating your paper, your instructor will consider the extent to which you:

- Identified the most important arguments needed to support your position. (When arguing for the privatization of prisons, did you cite examples of how this has worked efficiently in history?)
- Provided facts and information, when appropriate. (When arguing that prison expenditures are too high, you should state the actual amounts of recent expenditures and how the money was used in a wasteful manner.)
- Introduced new or creative arguments to those traditionally made on this issue. (Have new developments in technology enabled private business?)
- Presented your case accurately, coherently, logically, consistently, and clearly?

Conclude Concisely

Your concluding paragraph should sum up your argument clearly, persuasively, and concisely. When writing this assignment, follow the format directions in Chapter 3 of this manual. Ask your instructor for directions concerning the length of the paper, but in the absence of further directions, your paper should not exceed five (typed, double-spaced) pages.

The following sample reaction paper was selected from a criminal justice class instructed by Professor Rettig at the University of Central Oklahoma. Read it and assess its strengths and weaknesses. How well does it meet the criteria outlined above?

ARTICLE CRITIQUES

An article critique evaluates an article published in an academic journal. A good critique tells the reader what point the article is trying to make and how convincingly it makes this point. Writing an article critique achieves three purposes: First, it provides you with an understanding of the information contained in a scholarly article and a familiarity with other information written on the same topic. Second, it provides an opportunity to apply and develop your critical thinking skills as you attempt to critically evaluate the work of a criminal justice professional. Third, it helps you to improve your own writing skills as you attempt to describe the selected article's strengths and weaknesses so that your readers can clearly understand them.

Choosing an Article

The first step in writing an article critique is to select an appropriate article. Unless your instructor specifies otherwise, select an article from a scholarly journal (such as *Justice Quarterly, Criminology, Justice Professional,* or *Journal of Criminal Justice*)

A Negative Response to the Reaction Statement:

"Prisons Should Be Run by the Private Sector."

by

Scott M. Houck

for

Innovations in Corrections 4773
Section 3623

Professor Richard Rettig

University of Central Oklahoma

January 1998

Privatization of prisons is a mistake. Almost all private-sector industries and corporations are motivated toward the goal of making a profit. Private-sector management is mostly, if not totally, concerned with reducing overhead and increasing profit margin. In the prison setting, it would be too much of a temptation to cut overhead to such an extent that it violates the civil rights of offenders and/or threatens the health and safety of employees, offenders, and the community.

Although the current system is certainly not perfect, government-run prisons are not designed to make a profit for an individual or a group of individuals. A tax cut to the constituency is the main selling point of privatization. However, no matter what you do, it will still cost a lot of money to keep a person incarcerated. That inmate is still entitled to certain rights and liberties that have been granted by the Constitution and upheld in various court decisions. In general, the provisions of many of the rights granted by the Constitution require some form of funding. A profit-based company will test the very limits of humane incarceration by cutting costs on primary and secondary services and necessities provided to inmates. Services such as health programs, food necessities (quantitative and qualitative), treatment programs, vocational training, education, and religious provisions would be cut to increase the profit margin.

In the never-ending struggle for increased profits, personnel expenditures are the first items to be cut back. The result is lower salaries for entry-level employees, lack of or decreased in-service training events, decreased promotional opportunities, short staffing of shifts, increased potential for sick leave abuse, lack of or decreased health and retirement benefits, and reductions in other various personnel benefits.

The problem will begin in the recruitment process. Currently, it is very difficult to recruit qualified people into this thankless profession while paying a modest salary. The correctional officer position is one of the least respected, most underpaid, and most important job classifications in the prison environment. Not only must he supervise offender activity, maintain facility security, update or create reports, communicate with offenders, and defuse potentially violent situations but also try to maintain some form of a family and personal life. It is very difficult to maintain a decent family life when shift work constantly interferes with holidays, birthdays, weekends, and anniversaries. Compound that with minimal financial compensation that for this often unrewarding job and it's easy to see the difficulty in attracting qualified people. Private-sector prison management's concern with profit will make prison life even more demanding on everyone by employing just about anyone who applies for a job for as little as possible.

Another major consideration or argument against privatization of prisons is health care provisions. Due to the fact that a privatized prison is a profit-based organization, it is very likely that basic and preventative health care would be drastically reduced. Such procedures or items that are unnecessary to the survival of the offender would not be implemented. If they were provided, the cost would certainly be passed back to the taxpayer. Usually, if not always, when a contract is written by a private company with the government, certain items are included in the basic contract for the housing of the offender, such as, bed space, clothing, sundry items, and food. Other items that are not included, but provided at the cost of the state, are things like major surgery, kidney dialysis, some minor surgeries, emergency room costs, and other medical items that possibly come up during a person's incarceration. These "extras" add up quickly to be a major expense, one that is not included in the original contract.

There is much more to privatization of prisons than what meets the eye. On the surface it seems to sound great. However, once one reads "the fine print" of private-sector prisons, it will probably cost the taxpayer more in the long run, while jeopardizing the rights of inmates. Privatization of prisons is a mistake; no one but the company will benefit!

and not a popular or journalistic publications (such as *Time* or *National Review*). Your instructor may also accept appropriate articles from academic journals in other disciplines, such as history, political science, or sociology.

Three other considerations should guide your choice of an article: First, browse article titles until you find a topic that interests you. Writing a critique will be much more satisfying if you have an interest in the topic. Hundreds of interesting journal articles are published every year. The following articles, for example, appeared in a 1997 issue (volume 10, number 1) of *Justice Professional:*

- "Scamming: An Ethnographic Study of Workplace Crime in the Retail Food Industry"

- "A Writing-Intensive Approach to Criminal Justice Education"
- "Future Trends in Terrorism"
- "Problem-Oriented Policing: Assessing the Process"
- "The Legal Ramifications of Student Internships"
- "Retiring from Police Service: Education Needs and Second Career Planning"

The second consideration in selecting an article is your current level of knowledge. Many criminal justice studies, for example, employ sophisticated statistical techniques. You may be better prepared to evaluate them if you have studied statistics.

The third consideration is to select a current article, one written within the twelve months prior to making your selection. Much of the material in criminal justice is quickly superseded by new studies. Selecting a recent study will help ensure that you will be engaged in an up-to-date discussion of your topic.

Writing the Critique

Once you have selected and carefully read your article, you may begin to write your critique, which should cover the following four areas.

- Thesis
- Methods
- Evidence
- Evaluation.

THESIS

Your first task is to find and clearly state the thesis of the article. The thesis is the main point the article is trying to make. In an article selected from a 1997 issue (volume 10, number 1) of *Justice Professional,* Professors Michael Doyle and Robert Meadows of California Lutheran University's Department of Sociology and Criminal Justice examine "A Writing-Intensive Approach to Criminal Justice Education: The California Lutheran University Model." In this article, coincidentally on the importance of helping criminal justice students become better thinkers through writing, Doyle and Meadows state their thesis very clearly:

> The purpose of Criminal Justice education is to develop in students the knowledge, judgment, values, and ethical consciousness essential to becoming responsible citizens and leaders in the Criminal Justice system. Equally important is preparing students to critically evaluate and analyze justice issues through a variety of writing assignments. . . . By exposing students to a number of reflective, documentary, and analytical writing assignments, a better understanding of the justice process is achieved. (P. 19)

Sometimes the thesis is more difficult to ascertain. Do you have to hunt for the thesis of the article? Comment about the clarity of the author's thesis presentation, and state the author's thesis in your own paper. Before proceeding with the remaining elements of your paper, consider the importance of the topic. Has the author of the article written something that is important for criminal justice students or professionals to read?

METHODS

What methods did the author use to investigate the topic? In other words, how did the author go about supporting the thesis? In your critique, carefully answer the following two questions. First, were appropriate methods used? In other words, did the author's approach to supporting the thesis make sense? Second, did the author employ the selected methods correctly? Did you discover any errors in the way he or she conducted the research?

EVIDENCE

In your critique, answer the following questions: What evidence did the author present in support of the thesis? What are the strengths of the evidence presented by the author? What are the weaknesses of the evidence presented? On balance, how well did the author support the thesis?

EVALUATION

In this section, summarize your evaluation of the article. Tell your readers several things. Who will benefit from reading this article? What will the benefit be? How important and extensive is that benefit? What is your evaluation of the article? What suggestions do you have for repeating this study or one like it? Your evaluation might begin like the following:

> Doyle and Meadows' article on "A Writing-Intensive Approach to Criminal Justice Education: The California Lutheran University Model," is an excellent presentation on both the need for criminal justice students to become better writers, and how a variety of writing assignments in the curriculum can enhance a "better understanding of the justice process." If they wish to ensure that students are able to "critically evaluate and analyze" important issues, those involved in criminal justice education would be well advised to consider adopting this model.

When writing this assignment, follow the directions for formats in chapter 3 of this manual. Ask your instructor for directions concerning the length of the paper, but in the absence of further directions, your paper should not exceed five (typed, double-spaced) pages.

The sample article critique that follows was written by a student at the University of Central Oklahoma. As you read it, ask yourself how well this student followed the guidelines described above.

BOOK REVIEWS

Objectives of a Book Review

Successful book reviewers answer two questions for their readers:

- What is the book trying to do?
- How well is it doing it?

People who read a criminal justice book review want to know if a particular book is worth reading, for their own particular purposes, before buying or beginning to read it. These potential book readers want to know what a book is about, and the book's strengths and weaknesses, and they want to gain this information as easily and quickly as possible.

Your goal in writing a book review, therefore, is to help people decide efficiently whether to buy or read a book. Your immediate objectives may be to please your instructor and get a good grade, but these objectives are most likely to be met if you focus on a book review's audience: people who want help in selecting books to read. In the process of writing a review according to the guidelines given in this chapter, you will also learn about the following:

- the book you are reviewing and its content
- professional standards for book reviews in criminal justice
- the essential steps to reviewing books that apply in any academic discipline.

This final objective, learning to review a book properly, has more applications than you may at first imagine. First, it helps you to focus quickly on the essential elements of a book, to draw from a book its informational value for yourself and others. Some of the most successful professional and business people speed-read many books. They read these books less for enjoyment than to assimilate knowledge quickly. These readers then apply this knowledge to substantial advantage in their professions. It is normally not wise to speed-read a book you are reviewing because you are unlikely to gain enough information to evaluate the book's qualities fairly. However, writing book reviews helps you to become proficient in quickly locating the book's most valuable information and paring away material that is of secondary importance. The ability to make such discriminations is of fundamental importance to academic and professional success.

In addition, writing book reviews for publication allows you to participate in the discussions of the broader intellectual and professional community of which you are a part. People in law, medicine, teaching, engineering, administration, and other fields are frequently asked to write reviews of books to help others in their profession assess the value of newly released publications.

Elements of a Book Review

Book reviews in criminal justice and the social sciences contain the same essential elements of all book reviews. Since social science is nonfiction, book reviews within the disciplines focus less on writing style and more on content and method than reviews of works of fiction. Your book review should generally contain four basic elements, though not always in this order:

1. Enticement
2. Examination
3. Elucidation
4. Evaluation.

ENTICEMENT

The first sentence should entice people to read your review. Criminal justice studies do not have to be dull. Start your review with a sentence that both sums up the objective of the book and catches the reader's eye. Be sure, however, that your opening statement is an accurate a portrayal of the book as well as an enticement to the reader.

EXAMINATION

Your book review should encourage the reader to join you in examining the book. Tell the reader what the book is about. When you review a book, write about what is actually in the book, not what you think is probably there or what ought to be there. Do not

Critique

of

Johnson, J.D., N. E. Noel and J. Sutter-Hernandez. 2000. "Alcohol and Male Acceptance of Sexual Aggression: The Role of Perceptual Ambiguity." *Journal of Applied Social Psychology* 30(6): 1186–1200.

by

Nancy K. Hamilton

for

Criminology 3633

Section 4651

Professor Shawna E. Cleary

University of Central Oklahoma

April 10, 2001

THESIS

Studies have established that alcohol use disrupts cognitive functions, and the fact that alcohol intoxication is a factor in a significant proportion of "date rape" incidents has also been well established by prior research. This study attempts to document the relationship between the level of alcohol consumption and male acceptance of sexual aggression.

The researchers examined how the interpretation of subtle vs. explicit behavioral cues, as related to perceived sexual intent, is impacted by blood alcohol level. In this study, it was expected that the effects of alcohol consumption would be moderated by the behavior of the female, i.e. that when the female appeared receptive to the sexual advances, intoxicated male subjects would consider sexual aggression to be more acceptable than would sober subjects. If the thesis were correct, sexual aggression would be unacceptable even to intoxicated subjects, provided the female's behavior indicated a clear and consistent message of disinterest.

METHODS

Researchers recruited 118 volunteers through posters and class announcements at a medium-sized southeastern state university. The study participants included university staff and students, students' friends, and students' relatives. The study excluded those volunteers with self-reported alcohol/drug problems, alcohol-related arrests (other than one charge of DUI), and those with significant medical or psychiatric problems. Study participants were tested for blood alcohol level (BAL) at the beginning of the session and only those with an initial BAL of 0.00% were included in the study results. To minimize the probability of "demand bias," participants were told the experiment was intended to measure the effect of alcohol on visual acuity and social perceptions.

Subjects were given one of four beverages on the basis of random selection. Those in the control group received ice water, and knew they were part of the control group. The placebo group was given tonic water with 0.08-ml alcohol/kg body weight. The low-dose group received 0.33-ml alcohol/kg body weight, and the moderate-dose group received 0.75-ml alcohol/kg body weight. Drinks were served in such a way that participants, other than those in the control group, had no way of knowing what dosage they had received. Subjects were allowed 20 minutes to consume the beverage and then spent 25 minutes completing a visual-acuity task. At that point, the subjects were asked to participate in two social perception experiments, one of which was the videotaped interaction of a male/female couple at the beginning of a "blind date." Subjects were shown two interactions, one in which the female was enthusiastic about the upcoming date, touched the male's arm, and laughed extensively. In the second scenario, the female maintained a rigid posture, frequently checked her watch, and reminded the male of her need to end the date at the prescribed time. Both sequences ended as the couple left for the movies. Participants were than asked three questions: Should the man try to have sex with the woman, even if it means using force? Would you try to have sex with the woman, even if it meant using force? How responsible would the female be if the male forced her to have sex? Responses to the third question were rated on a 9-point scale ranging from 1 (not responsible at all) to 9 (totally responsible).

The authors of this study were diligent in structuring the experiment in a way that minimized unintended influences: 1) care was taken to ensure that participants were not influenced by their own or the researchers' expectations; 2) there were clear distinctions between the behaviors exhibited by the females in the two scenarios; and 3) variations in the experimental treatments (alcohol dosage) were consistent across experimental groups.

EVIDENCE

Researchers found a strong positive correlation between increased alcohol consumption, acceptance of sexual aggression, and attribution of responsibility to the female. As expected, study participants consistently rejected sexual aggression toward the female in the unreceptive scenario, but accepted sexual aggression toward the female in the receptive scenario, and were more accepting of that aggression as they become more intoxicated. Responsibility for the aggression was assigned to the female in the receptive scenario, to the male in the unreceptive scenario. These findings supported the thesis that unambiguous behavioral cues would be recognized and accepted even when the males were intoxicated, but that in a state of intoxication the males would attned to the most obvious behavioral cues and be more inclined to disregard other inhibitory cues (such as legal and moral sanctions against sexual aggression).

Participants' responses to the three questions indicated that BAL and the acceptance of sexual aggression increased in tandem, as did the attribution of female responsibility for the aggression, and those responses support the thesis of the experiment.

EVALUATION

This study is valuable to anyone who has occasion to be in social situations where alcohol consumption occurs. We know that males are influenced by the effects of alcohol in their interpretation of female behavior that appears to be sexually receptive. In the end, when a woman says no, it should always mean no. However, it is important to understand that this message can be obscured in a haze of intoxication from alcohol. Consequently, women need a clear understanding of how their actions may be perceived and interpreted by male companions who are under the influence of alcohol. And men need to know how alcohol influences their own perception of

> female sexual receptiveness, and the criminal consequences of sex-
>
> ually aggressive behavior when a woman says no.
>
> In continuing the pursuit of this and related research, it would
>
> be instructive to see if these findings apply across broader eco-
>
> nomic and educational lines, and to study how female intoxication
>
> impacts the perception of sexual aggression and the attribution of
>
> responsibility for that aggression.

tell how you would have written the book, but tell instead how the author wrote it. Describe the book in clear, objective terms. Include enough about the content to identify for the reader the major points that the author is trying to make.

ELUCIDATION

Elucidate, or clarify, the book's value and contribution to sociology by defining (1) what the author is attempting to do, and (2) how the author's work fits within current similar efforts in the discipline of sociology or scholarly inquiry in general. The elucidation portion of book reviews often provides additional information about the author. How would your understanding of a book be changed, for example, if you knew that its author is a leader in the feminist movement? Include in your book review information about the author that helps the reader understand how this book fits within the broader concerns of social science.

EVALUATION

After your reader understands what the book is attempting to do, she or he will want to know the extent to which the book has succeeded. To effectively evaluate a book, you should establish evaluation criteria and then compare the book's content to those criteria. You do not need to define your criteria specifically in your review, but they should be evident to the reader. The criteria will vary according to the book you are reviewing, and you may discuss them in any order that is helpful to the reader. Consider including the following among the criteria that you establish for your book review:

- How important is the subject matter to the study of culture and society?
- How complete and thorough is the author's coverage of her or his subject?
- How carefully is the author's analysis constructed?
- What are the strengths and limitations of the author's methodology?

- What is the quality of the writing in the book? Is the writing clear, precise, and interesting?
- How does this book compare with other books written on the same subject?
- What contribution does this book make to sociology?
- Who will enjoy or benefit from this book?

When giving your evaluation according to these criteria, be specific. If you write "This is a good book; I liked it very much," you have told the reader nothing of interest or value. But if you say, for example, "Smith's book provides adequate descriptions of most major theories in juvenile delinquency, but it fails to describe the full extent of Sutherland's theory of differential association," then you have given your reader some concrete information.

Types of Book Reviews: Reflective and Analytical

Two types of book reviews are normally assigned by instructors in the humanities and social sciences: the reflective and the analytical. Ask your instructor which type of book review she or he wants you to write. The purpose of a *reflective* book review is for the student reviewer to exercise creative analytical judgment without being influenced by the reviews of others. Reflective book reviews contain all the elements covered in this chapter—enticement, examination, elucidation, and evaluation—but they do not include the views of others who have also read the book.

Analytical book reviews contain all the information provided by reflective book reviews but add an analysis of the comments of other reviewers. The purpose is to review not only the book itself but also its reception in the professional community. To write an analytical book review, insert a review analysis section immediately after your summary of the book. To prepare this review analysis section, use the *Book Review Digest* and *Book Review Index* in the library to locate other reviews of the book that have been published in journals and other periodicals. As you read these reviews, use the following four steps:

1. List the criticisms (strengths and weaknesses) of the book found in these reviews.
2. Develop a concise summary of these criticisms, indicate the overall positive or negative tone of the reviews, and discuss some of the most frequent comments.
3. Evaluate the criticisms of the book found in these reviews. Are they basically accurate in their assessment of the book?
4. Write a review analysis of two pages or less that states and evaluates steps 2 and 3, and place it in your book review immediately after your summary of the book.

Format and Length of a Book Review

The directions for writing papers provided in Part 1 of this manual apply to book reviews as well. Unless your instructor gives you other specifications, a reflective book review should be three to five pages in length, and an analytical book review should be from five to seven pages. In either case, a brief, specific, concise book review is almost always preferred over one of greater length.

8 Criminal Justice Agency Case Studies

WHAT IS A CASE STUDY

A case study is an in-depth investigation of a social unit, such as a police department, a group of prisoners, or a juvenile detention agency, undertaken to identify the factors that influence the manner in which the unit functions. Some examples of case studies are:

- An evaluation of the comparative effectiveness of behavior modification methods in maximum-security facilities
- A study of management practices at Arkansas State Penitentiary
- A study of the personality types of juvenile justice officers.

Case studies have long been used in law schools, where students learn how the law develops by reading actual court case decisions. Business schools began to develop social service agency case studies to help students understand actual management situations. Courses in social organization, public administration, and social institutions adopt the case study method as a primary teaching tool less often than business or law schools, but case studies have become a common feature of many courses in these areas.

Psychologists have used the case histories of mental patients for many years to support or negate a particular theory. Criminologists and sociologists use the case study approach to describe and draw conclusions about a wide variety of subjects, such as labor unions, police departments, medical schools, gangs, public and private bureaucracies, religious groups, cities, and social class (Philliber, Schwab, and Sloss 1980:64). The success of this type of research depends heavily on the open-mindedness, sensitivity, insights, and integrative abilities of the investigator. Case studies fulfill many educational objectives in the social sciences. As a student in a criminal justice course, you may write a case study in order to improve your ability to:

- Analyze information carefully and objectively
- Solve problems effectively
- Present your ideas in clear written form, directed to a specific audience.

In addition, writing a case study allows you to discover some of the problems you will face if you become involved in an actual social situation that parallels your case study.

For example, writing a case study can help you understand:

- Some of the potentials and problems of society in general
- The operation of a particular cultural, ethnic, political, economic, or religious group
- The development of a particular problem, such as crime, alcoholism, or violence, within a group
- The interrelationships, within a particular setting, of people, structures, rules, politics, relationship styles, and many other factors.

USING CASE STUDIES IN RESEARCH

Isaac and Michael (1981:48) suggest that case studies offer several advantages to the investigator. For one thing, they provide useful background information for researchers planning a major investigation in the social sciences. Case studies often suggest fruitful hypotheses for further study, and they provide specific examples by which to test general theories. Philliber et al. (1980:64) believe that through the intensive investigation of only one case the researcher can gain more depth and detail than might be possible by briefly examining many cases. Also, the depth of focus in a single case allows investigators to recognize certain aspects of the object being studied that would otherwise go unobserved. For example, Becker et al. (1961) noticed that medical students tend to develop a slang, which the researchers called "native language." Only after observing the behavior of the students for several weeks were the researchers able to determine that the slang word *crocks* referred to those patients who were of no help to the students professionally because they did not have an observable disease. The medical students felt the crocks were robbing them of their important time.

Bouma and Atkinson (1995:110–14) call attention to the exploratory nature of some case studies. Researchers, for example, may be interested in what is happening within a juvenile detention center. Before beginning the project, they may not know enough about what they will find in order to formulate testable hypotheses. The researchers' purpose in doing a case study may be to gather as much information as possible in order to help in the formulation of relevant hypotheses. Or the researchers may intend simply to observe and describe all that is happening within the case being studied. Or as is the case in our juvenile detention study in this chapter, the intention may be to observe, describe, and measure certain behaviors in order to test predetermined hypotheses statistically.

LIMITATIONS OF THE CASE STUDY METHOD

Before writing a case study you should be aware of the limitations of the methods you will be using in order to avoid drawing conclusions that are not justified by the knowledge you acquire. First, case studies are relatively subjective exercises. When you write a case study, you select the facts and arrange them into patterns from which you may draw conclusions. The quality of the case study will depend largely upon the quality of the facts you select and the way in which you interpret those facts.

A second potential liability to the method is that every case study, no matter how well written, is in some sense an oversimplification of the events that are described and the environment within which those events take place. To simplify an event or series of events makes it easier to understand but at the same time distorts its effect and importance. It can always be argued that the results of any case study are peculiar to that one case and, therefore, offer little as a rationale for a general explanation or prediction (Philliber et al. 1980:65). A third caution about case studies pertains strictly to their use as a learning tool in the classroom: Remember that any interpretations you come up with for a case study in your class, no matter how astute or sincere, are essentially parts of an academic exercise and therefore may not be applicable in an actual situation.

TYPES OF CASE STUDIES WRITTEN IN CRIMINAL JUSTICE

Criminal justice cases usually take one of two basic forms. The first might be called a *didactic case study,* because it is written for use in a classroom. It describes a situation or a problem in a certain setting but performs no analysis and draws no conclusions. Instead, a didactic case study normally lists questions for the students to consider and then answer, either individually or in class discussion. This sort of case study allows the teacher to evaluate student analysis skills, and if the case is discussed in class, to give students an opportunity to compare ideas with other students.

The second form, an *analytical case study,* provides not only a description but an analysis of the case as well. This is the form of case study most often assigned in a criminal justice class, and it is the form described in detail in this chapter.

Criminal justice professionals conduct case studies for a variety of specific purposes. An *ethnographic case study,* for example, is an in-depth examination of a group of people or an organization over time. Its major purpose is to lead the researchers to a better understanding of human behavior through observations of the interweaving of people, events, conditions, and means in natural settings or subcultures.

Ethnographic case studies examine behavior in a community or, in the case of some technologically primitive societies, an entire society. The term *ethnography* means "a portrait of a people," and the ethnographic approach was historically an anthropological tool for describing societies whose cultural evolution was very primitive when compared to the "civilized" world (Hunter and Whitten 1976:147). Anthropologists would sometimes live within the society under scrutiny for several months or even years, interviewing and observing the people being studied. The in-the-field nature of ethnographies has caused them to be referred to occasionally as field studies.

HOW TO CONDUCT A CRIMINAL JUSTICE AGENCY CASE STUDY

Unlike an ethnographic study, which looks at a community or a society as a whole, a *service agency case study* focuses on a formal organization that provides a specific service or set of services either to a section of a society or society as a whole. A service agency case study usually describes and explains some aspect of a service agency's operation. Case studies do not attempt to explain everything there is to know about the organization. To conduct a service agency case study, you undertake the following tasks:

1. Select a particular service agency to study.
2. Formulate a general goal, for example, to better understand how the agency works.
3. Describe in general terms the agency and how it operates.
4. Describe the structure, practices, and procedures of the agency.
5. Select a specific objective, for example, to discover the responsiveness to the agency's programs of the people the agency serves.
6. Describe your methodology, that is, the procedures you will use to conduct your investigation.
7. Describe the results of your study, the observations you have made.
8. Draw conclusions about your findings.

The specific goal of your case study is to explain the effectiveness or ineffectiveness of some aspect of the agency's operation. This focused inquiry will, in turn, contribute to a general understanding of how the selected agency works. You may select any service agency, such as the United Way, the Social Security Administration, the American Red Cross, or your state's Department of Human Services. Its personnel, however, should be directly accessible to you for interviews.

Once you have chosen a service agency to examine, you then need to focus on a specific topic related to the agency—some characteristic procedure or situation—and write a description of how that topic has developed within the agency. For example, if your agency focus is a county health department and your topic focus is recruitment problems, you might chose to describe how the recruitment problems evolved within the overall operations of the department.

Most social service agency case studies assigned in criminal justice classes are not fictional. They are based on your investigation of an actual current or recent situation in a public or private agency.

Selecting a Topic

In seeking a topic, you are looking for a situation that is likely to provide some interesting insights about how social service agencies affect people's lives. There are two ways to begin your search. The first is to contact an agency involved with a matter that interests you and then inquire about recent events. If, for example, you are interested in the court system, you would contact a local court administrator's office. Tell the secretary who answers the phone that you are a student who wants to write a college term paper about the agency and ask to talk to someone who can explain the organization's current programs. Ask for an appointment for an interview with the person to whom you are referred by the secretary. When you arrive for the interview, tell the agency official to whom you are speaking that you are interested in doing a case study on some aspect of the agency's operations and that your purpose is to understand better how government agencies operate. Then ask a series of questions aimed at helping you find a topic to pursue. These questions might include:

- What recent successes has your agency had?
- What is the greatest challenge facing your agency at the moment?
- What are some of the agency's goals for this year?
- What are some of the obstacles to meeting these goals?

You should follow up these questions with others until you identify a situation in the agency appropriate for your study. There will probably be many. Consider the following examples:

- The agency faces budget cuts, and the director may have to decide among competing political pressures which services she must reduce.
- The agency faces a reorganization.
- The agency is criticized by the people it serves.
- The agency has initiated a controversial policy.

Another way to select a topic is to find an article of interest in your local newspaper. The successes, failures, challenges, and mistakes of government and private agencies are always in the news. The benefit of finding a topic in the newspaper is that when you contact the agency involved you will already have a subject to discuss. The disadvantage is that, on some publicized topics, agency officials may be reluctant to provide detailed information.

The Importance of Interviews

The goal of your first interview is to obtain enough information to request a series of other interviews. The answers to questions you pose in these interviews will allow you to understand the course of events and the agency interactions that have resulted in the situation you are studying. Remember that you are writing a story, but the story you are writing is accurate and factual. Do not accept the first version of a course of events that you hear. Ask several qualified people the same basic questions.

Take notes constantly. It's best to not use a tape recorder, because tape recorders tend to inhibit people from giving you as much information as they would without one present. At every interview ask about documents relevant to the case. These documents may include committee reports, meeting minutes, letters, or organizational rules and procedures. Sort out fact from appearance. When the facts are straight, you will be ready to organize your thoughts first into an outline and then into a first draft of your paper.

ELEMENTS OF THE CASE STUDY PAPER

Overview of Contents

Your case study should consist of four basic parts:

1. Title page
2. Executive summary
3. Text
4. Reference list.

The title page, text, executive summary, and references should all conform to the directions in Part 1 of this manual.

Text

The text of a service agency case study includes the following elements:

- Facts of the case
- Environment, context, and participants of the case
- Topic analysis
- Conclusions.

Although the content of the text should adhere to this general order, elements will overlap. Ask your instructor for the assigned length of the paper. In general, case studies should be brief and concise. They may include material from numerous interviews and documents—but only material essential to understand the case. A case study can be any length, but a paper of about fifteen pages, double-spaced, is usually adequate to describe and analyze a case situation accurately.

THE FACTS

The facts of the case that you will reveal include a description of the events, the major actors and their relationships with one another, and the external and internal agency environments and contexts within which the events of the situation you are describing developed.

THE ENVIRONMENT AND CONTEXT

In your account, consider the following aspects of a situation and relate to your readers those items that are relevant to the case at hand:

- The law under which the agency operates
- Political and economic factors of the agency's internal environment: power and influence, budget constraints, agency structure, rules, role, and mission
- Factors of the agency's internal environment: style, tone, preferences, and procedures.

Without altering the essential facts of the course of events, alter or delete the names of the actors and the agencies for which they work. Accuracy of facts in a case study is essential for a correct interpretation, but the actual identities of the individuals involved are irrelevant, and people may want their privacy protected. Any change of facts for this purpose should be done in a manner that does not alter the content of the story of the case at hand.

A well-written criminal justice agency case study will reveal much about how public and private agencies conduct business in the United States and even more about the agency selected for study. Public and private administrators face many of the same problems: They must recruit personnel, establish goals and objectives, account for expenditures, and abide by hundreds of rules and regulations. In several important respects, however, public agencies are very different from private businesses. A public administrator will often serve several bosses (governor, legislators) and have several competing clienteles (interest groups, the general public). Public officials are more susceptible than private businesses to changes in political administrations. They also face more legal constraints and are held accountable to higher ethical standards. In addition, public administrators are more likely to be held under the light of public surveillance, and so they are held accountable to a different bottom line.

The goal of most businesses is, first and foremost, to earn profits for their owners. The amount of these profits is normally easy to quantify. The success of public agencies, however, can be hard to measure. Criteria used to evaluate public programs, such as effectiveness and efficiency, often contradict one another. For example, the nation's space program has accomplished some remarkable achievements, but not many people commend the program's economic efficiency.

TOPIC ANALYSIS

Your analysis should explore and explain the events in your selected situation, concentrating upon the strategies and practices used by the primary actors at the social service agency. Your analysis should answer questions such as the following:

- How did the situation or problem at the heart of the case arise?
- What were the important external and internal factors that directed what transpired?
- What were the major sources of power and influence in the situation, and how were they used?
- What social service agency styles and practices were employed, and were they effective and appropriate within the situation described?
- How did relationships within the organization affect the conduct of other public or private programs?

CONCLUSIONS

To the fullest extent possible your conclusions should use what you have learned about the nature of administrative practices to explain causes and effects, summarize events and their results, and interpret the actions of administrators. One major purpose of a social service agency case study is to give its readers an opportunity to benefit from the successes and mistakes of others. In your conclusion, tell the reader what you have learned from this situation, what you would imitate in your own social service agency practice, and what you would do differently if you found yourself in a similar situation.

Summary

In this chapter we have introduced you to the importance of the case study in criminal justice. Completing a case study will help improve your ability to:

- Design and implement a plan to study a specific case
- Collect and analyze information carefully and objectively

- Solve problems effectively
- Present your ideas and recommendations in clear, written form, directed to a specific audience.

Doing a study of this type allows you to discover some of the issues you will face if you become involved in an actual social situation that parallels your case study. This mode of research provides the criminal justice student or professional with a powerful tool to describe and a wide variety of topics within the discipline.

Criminal Justice Policy Analysis Papers

WHAT IS POLICY ANALYSIS

Policy analysis is examining the components (analysis) of a decision to act according to a set principle or rule in a given set of circumstances (a policy). It is conducted at local, state, national, and international levels of government. Policy analysis is never completely technical; it is conducted within and immediately affected by numerous currents of political influence. The most publicized reports tend naturally to be the reports of Presidential Commissions, which are created by presidents to study possible government policies on a certain topic or problem and report their findings and recommendations.

Policy Analysis in Action:
The Challenge of Crime in a Free Society

Throughout the history of the United States there have been numerous presidential commissions, which have studied a wide range of subjects including crime, poverty, and violence. On July 23,1965, recognizing the urgency of the nation's crime problem and the depth of ignorance about it, President Johnson established the Commission on Law Enforcement and Administration of Justice through Executive Order 11236.

In the process of developing the findings and policy recommendations of the report, the Commission called three national conferences, conducted five national surveys, held hundreds of meetings, and interviewed tens of thousands of people. In its report, entitled *A Report by the President's Commission on Law Enforcement and Administration of Justice*, published in February 1967 by the Government Printing Office, the Commission made more than 200 specific recommendations, concrete steps the Commission believed could lead to a safer and more just society. These recommendations called for a greatly increased effort to enhance public safety on the part of the federal government, the states, counties, cities, civic organizations, religious institutions, business groups, and individual citizens. The recommendations called for basic changes in the operations of police, schools, prosecutors, employment agencies, defenders, social workers, prisons, housing authorities, and probation and parole officers. The central conclusion of the Commission was that a significant reduction in crime, while taking years to achieve, would be possible if the following objectives were vigorously pursued:

1. Society must seek to prevent crime before it happens by assuring all Americans a stake in the benefits and responsibilities of American life, by strengthening law enforcement, and by reducing criminal opportunities.

2. Society's aim of reducing crime would be better served if the system of criminal justice developed a far broader range of techniques with which to deal with individual offenders.

3. The system of criminal justice must eliminate existing injustices if it is to achieve its ideals and win the respect and cooperation of all citizens.

4. The system of criminal justice must attract more and better people— police, prosecutors, judges, defense attorneys, probation and parole officers, and corrections officials with more knowledge, expertise, initiative, and integrity,

5. There must be more operational and basic research into the problems of crime and criminal administration by those both within and without the system of criminal justice.

6. The police, courts, and correctional agencies must be given substantially greater amounts of money if they are to improve their ability to control crime.

7. Individual citizens, civic and business organizations, religious institutions, and all levels of government must take responsibility for planning and implementing the changes that must be made in the criminal justice system if crime is to be reduced.

In pursuit of these objectives, the Commission compiled over 200 specific policy recommendations. For example, society should undertake the following actions:

- Adopt centralized procedures in each city for handling crime reports from citizens, with controls to make those procedures effective
- Separate the present Index of Reported Crime into two wholly separate parts, one for crimes of violence and one for crimes against property
- Formulate police department guidelines for handling juveniles
- Provide alternatives to adjudication through Youth Services Bureau
- Divide court hearings for juveniles into adjudicatory and dispositional proceedings

- Establish community relations units in police departments serving substantial minority populations
- Clarify the statute authority of police to stop persons for questioning
- Adopt limiting use of firearms by officers
- Enact comprehensive state bail reform legislation
- Revise sentencing provisions of penal codes
- Improve university research and training in corrections
- Adopt state drug abuse control legislation
- Enact laws prohibiting transportation and possession of military-type weapons
- Require permit for possessing or carrying a handgun.

Well over one hundred of the Commission's specific policy recommendations, including all the above specific recommendations, have resulted in policy formation at the state and federal levels. Today these stand as accomplished laws, rules, and regulations for which specific policies and procedures have been put into place.

POLICY ANALYSIS RESEARCH PROPOSALS

Introduction to Policy Analysis Research Proposals

This chapter includes directions for two separate paper assignments. The first is a policy analysis research *proposal,* and the second is a policy analysis research *paper.* The proposal is a description of the research that will be conducted during the course of writing the research paper. The proposal assignment is included in this chapter because students who hope to become policy analysts will find that, in actual working situations, they will almost always be required to submit a proposal explaining and justifying the research that they expect to do before they are commissioned or funded to conduct the research itself. The proposal exercise that follows, therefore, prepares students for a task that they will face in the course of their careers when working in or consulting with public and private organizations.

The Purpose of Research Proposals

Research proposals are sales jobs. Their purpose is to sell the idea that a research study needs to be done. Before conducting a policy analysis research study for any public or private agency you will need to sell the job. This means that you will need to convince someone that:

- The study needs to be written
- The study will provide helpful information

- The study will be properly conducted
- You are qualified to conduct the study
- The cost of the study will be reasonable in comparison to the benefit it will provide.

Before a policy analysis study may begin, therefore, the person who will conduct the study must submit a *policy analysis research proposal* to the person who has authority to conduct the research. The purpose of this research proposal is to accomplish the following seven tasks:

1. Convince the people for whom the study is being done that the study is necessary
2. Describe the objectives of the study
3. Explain how the study will be done: the methods that will be used to conduct it
4. Describe the resources—time, people, equipment, facilities—that will be needed to do the job
5. Construct a project schedule which tells when the project will begin, when it will end, and important dates and times in between
6. Prepare a project budget which specifies the financial costs and the amount to be billed (if any) to the funding agency
7. Carefully define what the research project will produce, what kind of study will be conducted, its length and what it will contain.

The Content of Research Proposals

In form, policy analysis proposals usually contain the following four parts:

1. Title page
2. Outline page
3. Text
4. Reference page

In substance, policy analysis papers contain the information necessary to complete the tasks listed in the preceding section. The title page, outline page, and reference page should follow the guidelines defined in Chapters 3 and 4.

WRITING THE TEXT OF RESEARCH PROPOSALS: AN OVERVIEW

The text of a policy analysis research proposal contains the following seven elements:

1. An explanation of the *need* for the study
2. A description the *objectives* of the study
3. An explanation of the *methods* that will be used to conduct the study
4. A listing of the human, material, and financial *resources* needed to conduct the study
5. A project *schedule*
6. A project *budget*
7. A description of the anticipated *product* of the research project.

An Explanation for the Need of the Study. Have you ever listened to an automobile salesperson who is trying to sell a car to someone? The first thing that the salesperson will usually ask a customer, after getting the customer's name is, "What kind of a vehicle do you *need?*" Perhaps salesperson will follow up that question with, "How many people are in your family?" and then say something like, "It seems that you *need* a large vehicle." The good salesperson will not stop this line of questioning until the customer has agreed upon some statement of need. The first objective of a research proposal is to demonstrate that the people for whom it will be written *need* the information that the policy analysis will contain. They need the information because they are faced with one or more problems that their present policies are inadequate to handle.

Start your proposal, therefore, with a clearly written statement of need. Suppose that you believe, for example, that the current policy of the Springfield Board of Education is inadequate as it relates to predelinquent children. Your statement of need might be constructed like this:

> According to reports from school administrators, the Springfield Board of Education's policy on provision of supplemental services for predelinquent children (a policy that provides only for minimal services) leaves some district children without sufficient resources to complete their secondary education; this eventuates in an increase in dropout rates and the possible outcome of serious delinquency.

This statement clearly indicates a need for a review of the district's policy. Your need statement should be comprehensive enough to impress people that a definite need exists. It must be clear enough for people to understand the point immediately.

The Policy Problem and the Policy Deficiency

A *policy problem* is an actual problem that a public policy is supposed to solve. A policy problem is therefore a deficiency in a public policy that is caused by a problem independent of that policy. The *policy* of the Springfield Board of Education (minimal services) is deficient because it does not solve a *problem* (some children are assessed as predelinquent). In this case, the policy did not create the problem. The policy is deficient because it does not solve the actual problem. The policy problem in this case, therefore, is that some of the students are in trouble with the law. The policy *deficiency* is the fact that the current policy does not solve the problem of children in trouble, which in turn means that the district's goal of providing a secondary education to all district children cannot be achieved.

Statements of need in policy analysis proposals should clearly identify both a specific policy problem and a specific policy deficiency.

Research Imperatives

If a study is "imperative," it must be done. Not every need that an organization has will be called an imperative, because there are always more needs than can be met. Every successful organization, however, meets the needs that are imperative, and leaves some of the ones that are not imperative unfulfilled, at least temporarily. If your proposal merely states a need, the people to whom you are writing may decide that meeting the need you have identified is not imperative and may choose not to proceed with the study. A good proposal, therefore, includes a statement of research imperatives. This statement will impress people with the necessity of the proposed research.

To continue our example, policy imperatives for the problem of predelinquent children may include the idea that the school district, if it does not change its policy, may:

- Fail to meet its fundamental obligation to the children involved
- Fail to meet the community's need for educated, self-sufficient citizens
- Fail to meet state and federal standards for education.

When formulating a statement of policy imperatives, it is vitally important to follow two principles:

1. Make a strong case that the research is imperative.
2. Be completely accurate and honest: do not overstate your case.

DISCRIPTION OF THE STUDY OBJECTIVE. Think about the person for whom you are writing your research proposal. The first question this person will ask when presented with a research proposal, as we have already noted, is, "Why do I need this study?" After you have answered this question, that person will ask, "What will this study do for me?" If you were selling this person an automobile, you might say, "It will take you from 0 to 60 in 5.5 seconds!" or " It will take you, your five children, your dog and your parakeet to Altoona and back in complete comfort." Since you are selling policy analysis instead of automobiles, however, your answer to the question, "What will this study do for me?" will be, in general terms, "The proposed policy analysis research study will (1) more clearly define your problem, (2) identify deficiencies in your present policy, (3) examine different ways of overcoming these deficiencies, and (4) recommend the most promising solution."

Problem Clarification

Your policy analysis research *proposal* will include an initial definition and explanation of the problem, but since you have not yet investigated the policy problem in detail, the proposal will not provide a sufficiently clear picture of the character and extent of the problem being studied. Your proposal, therefore, will explain that the policy analysis study to be conducted will further clarify and examine the nature and size of the problem. In our continuing example, our proposal will state something like the following:

> The proposed policy analysis research study will clarify and quantify the policy problem: predelinquent children among district school children. The study will (1) determine the exact number of children in the district who experience some form of predelinquent labeling, and (2) determine the form and extent of delinquent behavior that each of these children has manifested.

Your proposal will now briefly explain how these tasks will be carried out. Using the above case, for example, you will want to explain whether the juvenile court will be consulted to determine the degree and extent of delinquent behavior or whether you will rely on school records to determine the status of each child.

Evaluation of Current Policy

Your proposal will then tell its readers that you will, if the research study is authorized, provide a thorough examination of current relevant policy. For our Springfield Board of Education, you will write something like:

> The proposed policy analysis research study will evaluate the extent to which the Springfield District's current policy meets and fails to meet the needs of children who have experienced the label of predelinquent.

You must now explain how you will carry out this task. Give the reader of the proposal a brief or general idea of what criteria you will use to evaluate the current policy.

Comparative Evaluation of Policy Alternatives

Because the current policy is inadequate, the major purpose of a policy research study is to examine different ways of solving the policy problem. If the present policy does not work, what other policies might do a better job? Your proposal will promise to conduct an evaluation of alternative policies. Following our example, you might say something like this:

> The proposed policy analysis research study will identify and evaluate alternatives to the present policy. It will list the comparative advantages and disadvantages of each policy option.

Follow-up this statement with a brief description of some of the alternatives that you, in the actual policy analysis study, may evaluate.

Presentation of Recommendations

Your policy analysis study may or may not actually recommend which option to choose. You determine whether or not to include a policy recommendation by asking the person or persons to whom the proposal is being submitted, before you submit it, if they want a recommendation included. Their answer will depend upon a number of social, political, or economic considerations specific to their particular circumstances. Many public policy issues are decided not on the basis of the technical merits but upon political or economic considerations. The decision of whether to build a stadium for sporting events or to build an auditorium for theater and ballet, for example, will probably depend more upon who supports sports and who supports ballet than upon the technical advantages of one facility over the other. Do not assume you know if a recommendation should be included. Always ask. If the answer is affirmative, you may state something like this:

> Based upon an examination of the comparative advantages and disadvantages of each option, the proposed study will recommend the adoption of a specific policy.

EXPLANATION OF THE STUDY'S METHODS. By this time the person or persons to whom you are submitting your proposal may be impressed with the precision, if not the length, of your answers. However, they will still ask, "How do you propose to do all this?" At this point, back in the showroom, our automobile salesperson has just been asked, "*How* does it take you from 0 to 60 in 5.5 seconds?" The automobile salesperson answers, "It's the fuel-injected V-12 engine." Like the automobile salesperson, you will need to answer the question, "How do you propose to do all this?" At this point you explain your *methodology,* the steps you will take in conducting your analysis. They will include, as a minimum, collecting and analyzing information and presenting it in a form that people can understand.

Research Process and Methods

Your research proposal will briefly describe the steps you will take to find, evaluate, and draw conclusions from the information that is pertinent to your study. The research process normally proceeds in three steps:

1. Data (information) collection: gathering the appropriate information
2. Data analysis: organizing the data and determining its meaning or implications
3. Data evaluation: determining exactly what conclusions may be drawn from the data

Your research proposal will: (1) state that you intend to carry out these three steps, and (2) explain briefly how you intend to do them. Returning to our Springfield Board of Education example, your research proposal will say something like this:

> The proposed policy analysis research study will (1) review the records of the district's children who have been labeled or classified as predelinquent, (2) using standards methods in social research, determine from these records how many children experience social and academic impairment and the extent of this impairment, and (3) using treatment perspectives recommended in the literature, determine the programs and equipment necessary to address the problems of each child.

Quality Control

Quality control is a formal procedure to ensure that a product meets all relevant standards and is free of defects. Quality control is important in development of any product, including a policy analysis report. In policy analysis it is normally provided by experts, people who have years of experience in dealing with the problem at hand. Your proposal should state (1) that quality control will be provided for the study, and (2) how it will be provided. Following our example, a proposal might say something like this:

> Dr. William Enright of Northwest Central University and Dr. Susan Bray of the American Juvenile Justice Association will provide quality control for the proposed study by reviewing the methods and results of the research.

LISTING THE RESOURCES NEEDED TO CONDUCT THE STUDY. Finally, the person who has commissioned your study will want to know how much money, time, and other resources the project will take. This section is most important in scientific or engineering studies where extensive experimentation or design work is carried out. For most policy analysis studies this section may be brief, although it should in some way address the material and describe the human resources that will be necessary to conduct the study. Following our Springfield example, your proposal may say:

> Conducting the proposed study will require the use of a computer with an integrated word processing data management software program. Paper for report production and reproduction equipment for making copies of the report will also be needed. The principal investigator possesses the necessary computer equipment and will contract for copy services. Conducting the study will require the following human resources:
>
> - A principal investigator familiar with policy and data analysis techniques
> - Two quality control advisers, one of whom is familiar with the standards of the American Juvenile Justice Association, and another who is familiar with policy analysis research methods
> - A research assistant who is capable of entering data from juvenile court records and school forms into the computer

PROJECT SCHEDULE. You should include a *research schedule* in the proposal which states:

- When the project will begin
- When the major phases of the project will begin and end
- When any preliminary, interim, or final reports will be issued
- When any special or particularly events in the research or analysis process will occur
- When the project will end.

PROJECT BUDGET. A project budget is the next section to be included in a research proposal. The budget will normally contain the following categories:

- Materials (paper, computer discs, supplies, etc.)
- Facilities (conference rooms or places with special capabilities)
- Equipment (laboratory equipment, copy machines, computers)
- Travel and other expenses
- Personnel.

For each category, list the item needed and its cost. The personnel section should list each person or position separately, the rate of pay, and the total amount per person.

DESCRIPTION OF THE ANTICIPATED PRODUCT OF THE RESEARCH PROJECT. The final section of the proposal will describe the anticipated product of your study. In other words, you tell the person or persons to whom you are writing the proposal exactly what they will receive when the project is done. If you are writing this paper for a class in social or criminal justice policy analysis, you will probably write something like the following:

> The final product will be a policy analysis research study from 25 to 30 pages in length, which will provide an analysis of the policy problem and an evaluation of alternative new policies that may solve the problem.

CRIMINAL JUSTICE POLICY ANALYSIS PAPERS

Definition of a Policy Analysis Paper

A policy analysis paper evaluates a decision. It reviews current and potential social policies It is a document written to help decision makers select the best policy to solve a particular problem. In writing a policy analysis paper, the author

- Selects and clearly defines a specific policy
- Carefully defines the social, governmental, economic, or other problem the policy is designed to solve
- Describes the economic, social, and political environments in which the problem arose and the existing policy was developed
- Evaluates the effectiveness of the current policy or lack of policy in dealing with the problem
- Identifies alternative policies that could be adopted to solve the selected problem, and *estimates* the economic, social, environmental, and political *cost and benefits* of each alternative policy
- Provides a summary comparison of all policies examined.

Policy analysis papers are written at all levels of government every day. Public officials are constantly challenged to initiate new policies or change old ones. They want to know how effective their current policy is, if they have a formal policy at all. They then want to know what options are available to them, what changes might they make to improve current policy, and what the consequences of those changes will be. Policies are reviewed under a number of circumstances. Policy analyses are sometimes conducted as part of the normal agency budgeting processes. They help decision makers decide what policies should be continued or discontinued. They may be very narrow in scope, such as deciding the hours of shift change at a juvenile detention center. Or they may be very broad in scope, such as deciding how the state will provide treatment of delinquents or protection for its citizens.

Purpose of Policy Analysis Papers

Successful policy analysis papers all share the same general purpose and the same general objective. The objective of a policy analysis paper is to inform policy makers about how public policy in a specific area of concern may be improved.

Public officials are employed full-time in the business of making public policy. Program directors at the state and national levels employ professional staff people and consultants who continually investigate bureaucratic policy issues and seek ways to improve standing policy. At the national level, the *Congressional Research Service* continually finds information for Representatives and Senators. Each committee of Congress employs staff members who help it review current laws and define options for making new ones. State legislatures also employ their own research agencies and committee staff. Legislators and other policy makers are also given policy information by hundreds of public interest groups and research organizations.

A policy analysis paper is a *practical* exercise. Its focus is neither theoretical nor general in nature. The object of the exercise is to identify and evaluate the policy options that are available on a selected specific topic of interest.

Contents of a Criminal Justice Policy Analysis Paper

SUMMARY OF THE CONTENTS

Criminal justice policy analysis papers contain six basic elements:

1. Title page
2. Executive summary
3. Table of contents, including a list of tables and illustrations
4. Text, or body. of the paper
5. References to sources of information
6. Appendixes.

PARAMETERS OF THE TEXT

Ask your instructor for the number of pages required for the policy analysis paper for the course you are taking. Policy analysis papers for undergraduate courses in social policy analysis often range from 20 to 50 (double-spaced, typed) pages in length.

Two general rules govern the amount of information presented in the body of the paper. First, content must be adequate to make a good policy evaluation. All the facts necessary for understanding the significant strengths and weaknesses of a policy and its alternatives must be included in the paper. If your paper omits a fact that is critical to the decision, that decision will very likely be a poor one.

Never omit important facts merely because they tend to support a perspective other than your own. It is your responsibility to present the facts as clearly as possible so as not to bias the evaluation in a particular direction.

The second guideline for determining the length of a policy analysis paper is to omit extraneous material. Include only the information that is helpful in making the particular decision at hand.

Format of a Policy Analysis Paper

TITLE PAGE

The title page for a policy analysis paper should follow the format provided in chapter 3 of this manual.

EXECUTIVE SUMMARY

The executive summary for a policy analysis paper should follow the format provided in chapter 3 of this manual. An executive summary is composed of carefully written sentences expressing the central concepts that are more fully explained in the text of the paper. The purpose of the summary is to allow the decision maker to understand, in as little time as possible, the major considerations to be discussed in the paper. Each statement in the summary must be clearly defined and carefully prepared so that the decision maker should be able to get a thorough and clear overview of the entire policy problem and the value and costs of available policy options by reading nothing but the summary.

The content of the executive summary follows the content of the text of the paper. A sample executive summary for a hypothetical policy analysis paper is presented below. A good way to structure the executive summary is to use the most important topic statements of the text as the sentences in the summary. Below is an example of an executive summary for a criminal justice policy analysis paper.

Provision of Supplemental Services for the Predelinquent

EXECUTIVE SUMMARY

The Springfield School District serves 18,000 students who attend three high schools, seven middle schools and twelve elementary schools. There are 223 elementary students who have been assessed as having significant predelinquent behavior requiring additional instructional resources. The current policy of the school district is to provide only minimal resources in response to this need. These resources include the part-time services of one special education teacher, who has had special training in suitable instructional techniques. A budget for counseling, library, and other instructional resources has increased from $250 per year in 1985 to $725 in 1993. This amount meets about 20 percent of present need, which includes reading materials, cassette tapes, and moral enhancement literature for teachers and counselors.

In recent years national disabilities assistance programs have raised social awareness, and teachers, administrators, and students are becoming more sensitive to the needs of students in trouble with the law. Budget cutbacks in the school system, however, have made it difficult even to approach needed materials and special teaching assistance. When the Air Force closed the Springfield Base last fall, the county unemployment rate rose to 11 percent, and real estate values dropped 14 percent. During the last election two conservative members were elected to the Board of Education who pledged to fight to reduce school operating expenses.

These troubled children are receiving only 15 percent of the total services needed. Special reading materials and skilled professionals are in short supply, limiting the opportunities for the children to learn. Children labeled predelinquent average about 2.7 years below reading ability and comprehension for their age. Three alternatives to the present policy of minimal assistance could be initiated to alleviate these problems. First, the school district could provide new additional funding for current activities by redirecting funds from other academic programs. The second alternative is to redirect funds from athletic programs to the special needs of the predelinquent. The third option is to redirect the current student volunteer activities program to raising revenue and direct tutoring services to designated students. The third alternative is more acceptable politically than the first two but would have less direct benefit.

TABLE OF CONTENTS

The table of contents of a policy analysis paper must follow the organization of the paper's text, in accordance with the format shown in chapter 3 of this manual.

TEXT OF A POLICY ANALYSIS PAPER

The text, or body, of the paper should contain five kinds of information:

1. Description of the policy to be analyzed
2. Description of the social, physical, economic, and political (including legal and institutional) environments in which the policy has been or will be developed
3. Evaluation of the effectiveness and efficiency of the applicable current policy
4. Evaluation of alternatives to the current policy
5. Summary comparison of policy options

A policy analysis paper should follow the outline shown below. Study this summary carefully.

OUTLINE OF THE CONTENTS OF A POLICY ANALYSIS PAPER

   I. Policy Description
      A. Clear, concise statement of the policy selected
      B. Brief history of the policy
      C. Description of the problem the policy was aimed at resolving, including an estimate of the extent and importance of the problem

II. Policy Environment
  A. Description of the social and physical environmental factors affecting origin, development, and implementation of the policy
  B. Description of the economic factors affecting origin, development, and implementation of the policy
  C. Description of the political factors affecting origin, development, and implementation of the policy
III. Effectiveness and Efficiency of the Current Policy
  A. Effectiveness: How well does the existing policy do what it was designed to do?
  B. Efficiency: How well does the policy perform in relation to the effort and resources committed to it?
IV. Policy Alternatives
  A. Possible alterations of the present policy, with estimated costs and benefits of each alteration
  B. Alternatives to the present policy with estimated costs and benefits of each alternative.
  V. Summary Comparison of Policy Options

POLICY DESCRIPTION. The first task of a policy paper is to describe the policy that currently exists. The purpose is to allow the reader to understand the present situation and how this situation developed. Policy descriptions contain three basic elements. The first is a clear and concise statement of the policy selected. You may find that there is no written policy on the subject under consideration. If this is the case, you should first inform your readers of the lack of a written policy that addresses the subject, and then list and describe any policies that may indirectly affect the problem you are addressing. For example, a school district may have no policy that explicitly addresses destructive gang activities, but it may have policies on acceptable attire in school, rules of order in the classroom, and so on.

Where written formal policies exist, quote them directly from the documents that state the policies and provide the source of the quotation. If the policy is established under the authority of more general legislation, quote and cite the legislation also. For example, the laws of Indiana may include a law that says, in part, "School districts may establish appropriate rules of behavior and guidelines for suspension or expulsion for violations of these rules." The public schools of the city of Indianapolis, then, may have a regulation that says, "Students shall not wear clothing that unduly distracts or offends other students." Further, a particular school principal may have a policy that states, "shirts without sleeves shall not be worn in class during school hours except when participating in athletic activities." In these three examples we find three levels of public policy, all leading to a specific guideline applied in a specific school. Preparing a complete description of the policy at hand is very important if analysis of the problem is to be accurate and beneficial.

The second part of this section of your policy analysis paper should include a brief history of the policy. Several basic questions should be answered here. When was the existing policy first initiated? How did it come about? In response to what problem or need did it come about? What effects did the policy have? Remember that the purpose of this narrative in the paper is to help the reader understand what the present policy is and how it came about.

The third part of this section is a description of the problem the policy was aimed at resolving, including an estimation of the extent and importance of the problem. Carefully define the original problem that gave rise to the present policy. Estimate the size and extent of this problem. If the policy being examined is a school's dress code, estimate the number of students who have violated the code and the seriousness of the violation. These estimates are crucial for proper policy formulation. For example, suppose that a school district has 18,000 students. If six students at one high school are wearing fluorescent shorts, then a relatively minor problem exists. If 200 students come to school wearing gang insignia, a vastly more difficult problem has arisen. You need to provide the reader with an accurate assessment of the extent of the problem.

POLICY ENVIRONMENT. No policy exists in a vacuum. Wardens, police chiefs, and program directors do not sit in ivory towers cut off from the real world, and if they act as if they do, they are likely to pay a heavy political cost. Policies almost always arise from genuine needs, but they often reflect the needs of certain part of the population more than others. Policy is formulated within a number of environments. The most important of these are:

1. *Social environment*: the cultural, ethnic, and religious habits, practices, expectations, and patterns of relating to one another that every society establishes
2. *Physical environment*: the climate, architecture, topography, natural resources, and other physical features that shape the patterns of life in a society
3. *Economic environment*: the content and vitality of economic life, including the type of industry and commerce, the relative wealth or poverty of the area affected by the policy, the unemployment rate, and the rate of economic growth
4. *Political environment*: the government structures, the applicable laws, the political parties, the prevalent ideologies of the people, and the salient issues of the day.

The policy paper should describe each of these environments, giving more emphasis to those that have greater effects upon the policy under consideration. A juvenile detention center's dress code policy, for example, may be much more responsive to the social and economic environments than to the physical or political environments in a community. In each case, describe the environmental factors that affect the policy. A dress code will probably be influenced by the socioeconomic class or homogeneity of the neighborhood the facility is in. The policy paper should include separate discussions of the social, physical, economic, and political environments affecting policy development, giving different amounts of attention to each one according to what is important to the particular policy being discussed.

POLICY EFFECTIVENESS AND EFFICIENCY

Policy Effectiveness

An evaluation of policy effectiveness should tell the reader how well a specific policy does what it is intended to do. There are many methods for evaluating effectiveness and many factors that could be used in the analysis. This chapter will not explain these methods, but they may be found in any basic text on policy analysis. Instead of explaining the wide array of policy evaluation methods, this chapter will describe the basic steps that must be taken to complete any evaluation. *Ask the course instructor for specific directions or methods for policy evaluation.* The general explanation provided here may or may not be sufficient for the analysis paper that you are assigned.

Evaluation of policy effectiveness proceeds in three steps:

1. Constructing an evaluation framework
2. Applying the framework to the policy being examined
3. Drawing conclusions from your application.

The first step in policy evaluation is to establish an evaluation framework. An evaluation framework starts with a clearly defined policy to be evaluated. It then adds to the policy definition (1) a list of the general goals the policy is designed to meet, (2) a list of specific objectives that lead to the goals, (3) a list of criteria for judging the extent to which the objectives are met, and (4) a list of specific measurements for quantifying the fulfillment of the criteria.

To construct an evaluation framework:

1. Define the *general goals* the policy is intended to meet. Goals are general statements of the ends for which policy is made. The goal of a policy limiting the number of continuous hours that may be worked by a correctional officer, for example, is to maintain security for inmates, staff, and citizens in the community. Sometimes goals have been defined by the policy-making body that issues the policy. If not, attempt through interviews or researched documents to determine the goals for which the policy was made.
2. Define *specific objectives* that indicate partial accomplishment of the goals. Objectives are specific steps toward reaching goals. If the goal of the National Association of Police Chiefs is to make citizens feel more secure, for example, an objective might be to improve community policing programs in urban areas.
3. Define specific *criteria* for determining the extent to which the objectives are met. Criteria help you know the extent to which objectives are met. If the objective of a policy is to improve community policing programs, then criteria for determining the quantity and quality of services related to neighborhood foot patrol and answering calls for help need to be determined. One criterion might be the increase in personal contacts with citizens. Another criterion might be reduction in criminal and delinquent activity. Still another might be the effectiveness and visibility of 24-hour neighhood foot patrol. Establishing the proper evaluation criteria is one of the most difficult and important tasks in writing a policy analysis paper.
4. Define specific *measurements* for applying the criteria to the objectives. Measurements determine quantities. If a criterion for good community policing programs is the quantity and quality of neighborhood foot patrol, then valid measurements of productive contacts might include (1) the number of contacts per hour, (2) the number of problems or incidents handled, or (3) the reduction of criminal or delinquent acts in the neighborhood.

After establishing your framework, use the measurements you have designed. Try to make as many quantitative measurements as the study will allow. You may also need to make qualitative evaluations of variables that are not easily quantifiable, e.g., citizens' feelings about the officers that patrol their neighborhood.

Your final step may now be taken: drawing conclusions from your measurements. A good general practice is to be conservative in your evaluations. This means that you should interpret the data to indicate only what the data clearly demonstrate—not tendencies or implications that still need more proof.

Efficiency

Efficiency goes a step beyond effectiveness. Efficiency relates cost to accomplishment. Efficiency concerns not only how well a job is done, but also the amount of resources committed to getting the job done. In other words, when one inquires about the efficiency of a policy, one is asking the question, "How well did the policy perform in relation to the effort and resources committed to it?" Fuel efficiency in an automobile is determined by dividing the number of miles traveled by the number of gallons of gasoline consumed. An automobile that travels 50 miles for every gallon of gas is considered highly efficient. A program to reduce the number of neighborhood crimes across the city may require the expenditure of $2 million. If it reduces the number of criminal or delinquent acts by one thousand, then the efficiency of the program may be expressed in terms of the cost-per-crime prevention ratio, or $2,000 per crime reduced.

Evaluating efficiency is often more difficult than it appears. The most important factor in producing a valid efficiency evaluation is selecting the proper factors to place in the ratio. For example, to evaluate the efficiency of a prison arts program, one may divide the number of dollars in the annual budget for this program (perhaps $25,000) by the number of inmates who complete the requirements of the program (say, 250). The resultant figure would be $100 per inmate. This figure does not, however, provide any information about the quality of education that the inmates received, or the effectiveness of the program in combating recidivism. It is, therefore, a limited measure of efficiency. When writing your policy analysis paper, try to construct an efficiency ratio for the measurements you have made. Be very careful, however, to point out the limitations of the ratio you construct.

POLICY ALTERNATIVES. As author of a policy analysis paper, you want to be effective in clearly analyzing a policy and in assisting a public official to improve it. Having defined the deficiencies in the present policy, the next question you will ask is, "What can I do about it?" One of the most common mistakes of students who write policy analysis papers is their failure to answer this question correctly. To define options does *not* mean to describe several different measures that are all combined in one approach, but instead to describe *different approaches* to solving the problem. *Options* in policy analysis papers are different, mutually exclusive approaches to a problem. If one alternative is selected, the others are rejected. Sometimes students mistake the steps necessary to complete one course of action for distinct, separate approaches to the problem.

If student Jan Smith is studying a prison overcrowding policy, she may list as alternative approaches:

Option A: Demonstrate present prison overcrowding.

Option B: Draw up plans for a new state prison.

Option C: Generate funds and state legislative support for the facility.

Option D: Construct the prison in three phases over a two-year period.

Do you see a problem here? All four "options" are actually only four steps in carrying out *one* option: to build a prison.

Again, *steps toward a single solution are not options*. An actual set of options for solving the problem of prison overcrowding might be:

Option A: Continue present policy of encouraging prerelease work centers.

Option B: Plan and construct a new prison.

Option C: Expand alternatives to incarceration.

Option D: Broaden parole and probation services.

Conducting a Simplified Benefit/Cost Analysis

For each available option the student should describe first the benefits and then the costs. *Benefits* are the positive outcomes expected. *Costs* include time, money, and resources expected, and also probable or possible negative outcomes (known as "disbenefits"). Both benefits and costs may include economic, social, political, and environmental factors. Benefits of improving the public penal system, for example, may include:

1. Less prison overcrowding
2. Safety of staff, inmates, and citizens in the community
3. Decreased criminal activity
4. Increased economic development
5. Increased community security.

Costs of improving the prison system may include:

1. Construction costs of several million dollars
2. Fewer resources for other community projects, such as common education
3. Political opposition from areas in the community that do not benefit directly.

As mentioned previously, all reasonable costs and benefits should be included in the paper. The student should never exclude possible costs or benefits of any one option in order to make another option appear more or less attractive. The policy analysis paper will be submitted to someone else who is responsible for making the proper decision. If an option other than the one recommended is selected and full information has been provided, the writer of the position paper has acted appropriately. Elected and appointed officials may legitimately choose options which are technically not the most cost-effective. But if a policy maker selects the wrong option, making a decision based upon incomplete information in the policy analysis paper, the writer of the position paper is clearly responsible.

Lengthy books have been written on how to conduct benefit/cost analyses. Only a simplified (but very helpful) process will be outlined here.

In order to analyze benefits and costs *for each option,* always proceed with the same set of steps. The first step is to list the benefits and then the costs of that option. The second step is to assign a monetary value to each cost and benefit. Value estimates should be made according to the advice of experts, who usually include government officials, engineers, or consultants, or members of private organizations with expertise in the area. Price estimates for construction costs may be obtained from engineers in local, state, or national government agencies. Personnel cost estimates may be determined by multiplying the number of person-hours or days by the cost per hour for the services desired.

Some costs are more difficult to estimate. Intangible items such as "political discontent" are not easily quantified, and yet an attempt may reasonably be made to do so. Public policy analysts call such estimates *compensating variations.*

A compensating variation allows analysts, for comparison purposes, to place a dollar value on intangible factors. The dollar value of an intangible cost is the amount that the average person would normally accept as fair compensation for having paid a particular cost or endured a particular disbenefit.

For example, suppose that Mrs. Williamson lives in a quiet residential neighborhood. A community treatment center is planned at the end of her street that will increase traffic, noise, and personal risk near her home. Mrs. Williamson is unhappy about the treatment center. If asked if she would prefer that it not be built, she would answer yes. Suppose, however, that Mrs. Williamson was offered this choice: either (1) no center will be built; or (2) the center will be built and Mrs. Williamson will be paid $100.00. Mrs. Williamson may still prefer alternative 1. Suppose that the offer was increased to $1,000.00. Mrs. Williamson might then choose to have the money and the treatment center rather than neither the money nor the center. The compensating variation in this case would be the amount that Mrs. Williamson would be willing to accept as fair compensation for the treatment center. The social cost of building the treatment center may then be calculated as the sum of the compensating variations of the residents to be adversely affected by the facility.

Remember that all such estimates can be held to the judgment of the people who will read the paper. Estimates need be neither exact nor perfect. It is important to realize that someone must make a decision about whether or not to make the corrections system improvements and that a reasonable estimate is more helpful than no estimate at all.

Economists and urban planners have long known that attitudes affect development potential. Beliefs about the strength of the economy, for example, have a direct daily effect upon stock exchanges. Popular discontent makes it less likely that people will find a particular neighborhood a desirable place to live. In order to estimate discontent as a compensating variation, a brief survey of community attitudes would be necessary. Telephone calls to the 15 homes closest to the planned treatment center would yield several interviews with homeowners. Asking a few simple open-ended questions would make it possible to estimate the strength of sentiment about the treatment center in the neighborhood. It makes a major difference to the study whether one or two people are mildly concerned or whether 15 families are planning to move immediately if the facility is built. Mild discontent might be assigned no economic value ($0), whereas widespread strong feelings may make the neighborhood so undesirable that property values decline by 25 percent (a total loss in property value of $328,125). Both extremes are unlikely in this case, but one can see that it is possible to put a ballpark estimate on popular discontent.

The second part of benefit/cost analysis in the paper is entered in the first section of the third part of your paper: Recommendations. Here, a summary of benefit/cost analyses for all alternatives is presented in table form, followed by a paragraph or two of explanatory comparisons. The table for Jan Smith's paper might look like the table below.

Presentation of Policy Analysis Papers

All sources of information in a policy analysis must be properly cited. Follow the directions in chapter 4 in this manual.

Appendixes can be helpful to the reader of policy analysis papers. They provide information that supplements the important facts contained in the text. For many local development and public works projects, a map and a diagram are often very helpful. The student should attach the appendixes to the end of the paper after the reference page. The student should not append entire government reports, journal articles, or other publications, but selected charts, graphs, or other material may be appended. The source of the information should always be evident on the appended pages.

Benefit/Cost Analysis Summary Table

Alternative	Benefit	Dollar Value	Cost	Dollar Value
Prerelease Center	Reduces prison population, moderate economic development	1.5M	Maintenance, increase in public anxiety	4K per inmate per year
New Prison	Containment, high economic development	14M	Maintenance, large outlay of tax dollars	Construction plus 25K inmate per per year
Broader Parole and Probation	Quick to install, low disruption, no new construction	2M	Maintenance, more ex-cons on the street	2M per year

Glossary

accountability The obligation to perform responsibly and exercise authority in terms of established performance standards.

accreditation The status achieved by a correctional program, police agency, court, or other facility when it is recognized as having met certain national standards following an on-site audit by the relevant commission on accreditation.

accusation A formal charge against a person inferring that he or she may be guilty of a punishable offense.

accusatory instrument As usually defined, an indictment, information, simplified traffic information, prosecutor's information, misdemeanor complaint, or felony complaint. Every accusatory instrument, regardless of the person designated therein as the accuser, constitutes an accusation on behalf of the state as plaintiff and must be entitled "the people of the state of . . ." against a designated person, known as the *defendant*.

accusatory stage In police practice, the investigation that occurs after suspicion has focused on one or more particular individuals as having guilty knowledge of the offense. Distinguished from the investigatory stage, during which the offense is the subject of general investigation before suspicion has focused on a particular accused. Also use to mean the stages in a criminal prosecution from arrest to conviction or acquittal.

accused The generic name for the defendant in a criminal case.

acquittal A legal and formal certification of the innocence of a person charged.

act of God An act occasioned exclusively by violence of nature without the interference of any human agency; an act, event, happening, or occurrence due to natural causes.

actus reus The criminal act; the act of a person committing a crime.

adaptative behaviors Various responses through which inmates adjust to the institutional setting, including such psychological defense mechanisms as rejecting authority, projecting blame, or rationalization.

addict Someone who has a physical and psychological dependence on one or more drug(s), who has built up a physical tolerance that results in taking increasingly larger doses, and who has an overpowering desire to continue taking the drug(s). One can become addicted to illegal as well as legal drugs (such as alcohol).

adjudicate   To determine finally; to adjudge.

adjudication   Giving or pronouncing a judgment or decree in a cause; also, the judgment given. The equivalent of "determination"; the formal finding of guilt or innocence by a court of law. The stage that would be considered "trial" in the criminal justice system, which in juvenile court refers to a hearing to establish the facts of the case.

administration   The act of administering, or the state of being administered. Management of direction of affairs; the total activity of a manager.

administrative remedies   Formal administrative mechanisms used within correctional institutions to proactively reduce litigation, such as the implementation of grievance procedures to identify and address complaints.

administrative segregation   Separate confinement for inmates who, for any of a number of reasons, need closer attention or supervision than is available in the general population.

administrative sentencing model   A sentencing plan in which legislatures and judges prescribe boundaries but administrative agencies determine the actual length of sentence.

adversary   An opponent; the opposite party in a writ or action.

adversary proceeding   One having opposing parties, as distinguished from an exparte proceeding.

adversary process   The view of criminal justice as a contest between the government and the individual.

adversary system   The practice of conducting a legal proceeding as a battle between opposing parties under the judge as an impartial umpire with the outcome determined by the pleading and evidence introduced into courts; in Anglo-American jurisprudence includes the presumption of innocence of the accused. To be distinguished from the accusatory system used in continental law, where the accusation is taken as evidence of guilt that must be disproved by the accused.

advocate   One who assists, defends, or pleads for another.

affidavit   A written or printed declaration or statement of facts, taken before an officer having authority to administer oaths.

affirmative action   Programs instituted by government to increase the number of minority employees in the public and private sectors.

aftercare   Follow-up services and supervision provided upon release from a juvenile correctional institution, similar to parole or supervised mandatory release in the adult system.

aggravated assault   Assault with intent to kill or for the purpose of inflicting severe bodily injury; assault with the use of a deadly weapon.

aggregate crime rates   The number of crimes per 100,000 of the general population.

aggregate data   Data on large numbers of subjects showing a common characteristic.

aggressive field investigation   An investigatory model in which police consistently check out suspicious circumstances, places, and persons.

alcoholism The disease associated with abuse of legally available drugs (such as beer, wine, and/or liquor). Tendencies toward alcoholism may be inherited genetically, as well as precipitated by social and psychological factors.

alias Otherwise, in another manner; a fictitious name.

alienist One who specializes in the study of mental disease.

allocution The court's inquiry of a prisoner as to whether he or she has any legal cause why judgment should not be pronounced against him or her upon conviction.

alternative dispute resolution A type of formal diversion in which a neutral third party attempts to reach an agreeable compromise between the victim and the offender to resolve the case outside the criminal justice system.

amicus curiae A friend of the court. Also, a person who has no right to appear in a suit but is allowed to introduce argument, authority, or evidence to protect his interest.

anomie The weakening of social norms, which has been linked with crime and delinquency.

anonymity The assurance that research subjects' identities will not be disclosed.

anthropology A discipline focusing on the nature of human culture, in which field research is the primary method of study.

appeal The removal of a case from a court of inferior jurisdiction to one of superior jurisdiction for the purpose of obtaining a review and retrial.

appearance The coming into court as party to a suit.

appearance ticket A written notice issued by a public servant, requiring a person to appear before a local criminal court in connection with an accusatory instrument to be filed against him therein.

appellant The party who takes an appeal from one court of justice to another. In criminal law, usually the defendant in the lower court.

appellant jurisdiction The right of a court to review the decision of a lower court; the power to hear cases appealed from a lower court.

appellee The party in a cause against whom an appeal is taken; in criminal law, usually the state or the United States.

applied research Research for which one of the primary purposes is that the study may have some practical use.

archival research A method of studying organizations or societies based on the collected records they have produced.

argot The unique vocabulary used in communication between inmates and street criminals.

arraign To bring a prisoner before the court to answer an indictment or information. In practice, used to refer to any appearance of the accused before a magistrate or before a trial court to enter a plea. *See also* arraignment.

arraignment The proceeding for arraignment of the accused at which he enters a plea to the charge.

arrest The taking of a person into custody to answer to a criminal charge. A detention of a suspect subject to investigation and prosecution.

arson The intentional and unlawful burning of property. At common law, the malicious burning of the dwelling or outhouse of another.

asportation The removal of things from one place to another, such as required in the offense of larceny in some states.

assault The intentional unlawful use of force by one person upon another. If severe bodily harm is inflicted or a weapon is used, the offense is *aggravated assault*. The lesser degree of the crime is called assault or *simple assault*.

assault and battery A battery is an unlawful touching of the person of another. *See also* assault.

assessment A screening procedure in which a candidate's strengths and weaknesses are evaluated by a team of trained assessors on the basis of performance.

assessment center The place where screening procedure or assessment is implemented.

atavistic Having the characteristics of savages, as in early forms of human evolution.

attrition The loss of members of a sample, usually as a result of their refusal to respond or the researcher's inability to contact them.

attrition of cases Cases dropped at various stages in the criminal process.

Auburn system The approach that focused on congregate work and harsh discipline, as practiced at the correctional facility in Auburn, New York.

authority The sum of the powers and rights assigned to a position, such as a chief of police or a warden.

auto theft Stealing or driving away and abandoning a motor vehicle. May exclude taking for temporary use, or the taking for temporary use may carry a smaller penalty.

back-door strategy Reducing a prison population by the early release of prisoners.

background forces Psychological, biological, and sociological causes of crime.

bail To procure the release of a person from legal custody by instructing that he or she must appear at the time and place designated and submit himself or herself to the jurisdiction and judgment of the court.

bail bond A bond executed by a defendant who has been arrested, together with other persons as sureties, naming the sheriff, constable, or marshal as obligee, to receive a court-specified sum on condition that the defendant must appear to answer the legal process.

bailee One to whom goods are delivered under a contract or agreement of bailment.

bailor One who delivers goods under a contract or agreement of bailment.

bailment A delivery of goods or personal property by one person to another to carry out a special purpose and redeliver the goods to the bailor.

banishment and exile (also called *transportation*) Forms of punishment in which offenders were transported from Europe to distant lands (such as the Americas and Australia).

bar graph A graph on which the categories of a variable are presented on the horizontal axis and their frequencies on the vertical axis. The height of each bar represents the frequency of each attribute of a variable. The bars have gaps between them on the scale.

basic research   Research whose primary purpose is to contribute to systematic knowledge in a discipline.

behavior modification   Changing behavior through the conditioning power of such reinforcements as rewards and punishments.

bench warrant   A process issued by the court itself, or "from the bench," for the attachment or arrest of a person; either in case of contempt, or whether an indictment has been found, or to bring in a witness who does not obey a subpoena. So called to distinguish it from a warrant issued by a justice of the peace or magistrate.

beyond a reasonable doubt   Proof to a moral certainty, satisfying the judgment and consciences of the jury, as reasonable men and women, that the crime charged has been committed by the defendant. Moral certainty is not a requirement in some states.

bivariate table   A two-variable table.

bond   Money or property required to obtain release from jail for a criminal charge in order to assure appearance at trial.

booking   Creating an administrative record of an arrest. The clerical process involving the entry on the police "blotter" or arrest book of the suspect's name, the time of the arrest, the offense charged, and the name of the arresting officer. Used in practice to refer to the police-station-house procedures that take place from arrest to the initial appearance of the accused before the magistrate.

brainstorming   The exploration, discovery, and development of details to be used in a research study.

breach of the peace   A violation or disturbance of the public tranquility and order.

breaking and entering   Any unlawful entry even if no force was used to gain entrance.

building-tender system   Where prisoners assist officers in managing cell blocks.

burglary   Breaking and entering with intent to commit a felony or theft, or in some states, with intent to commit any offense.

capability   The range, variety, and depth of skills that a person holds in a certain job or position in the organization. It is also the sum total of the structure, process, and systems that make up the organization itself.

capacity   The outside limits of a person or organization or population that determines the amount of accomplishment or production that can done. Capacity, for example, can be increased by adding resources, restructuring the application of resources, or increasing the capability of resources within the organization. In criminal justice, population capacity is usually a matter of law.

career criminals   Criminals who devote the greater part of their lives to crime.

career-criminal units   Prosecutors who specialize in the prosecution of repeat offenders.

carnal knowledge   Sexual intercourse; the slightest penetration of the sexual organ of the female by the sexual organ of the male.

carrying concealed weapons   All violations of regulations or status controlling the carrying, using, possessing, furnishing, and manufacturing of deadly weapons.

case A general term for an action, cause, suit, or controversy in law or equity; a question contested before a court of justice.

case studies Observational studies of a single environment (a detention center, a police precinct, a public place). Field research is often based on a single case study.

caseload The workload of a probation (or parole) officer, usually measured by the number of cases being supervised and/or investigated.

caseload classification (also called *case management*) Separating cases according to the intensity of supervision needed by the client.

causation One or more variables produces a result. *See also* independent variable.

cell In a cross-classification table, the position where two categories meet. In a 2 × 3 bivariate table, there will be six cells. Also, an inmate's living quarters.

cell searches (also called *shakedowns*) A thorough examination of the structure and contents of a cell to detect and remove any contraband items. Searches may be conducted routinely, randomly, or based on suspicion, but not for harassment.

central intake (often called a *diagnostic* or *reception center*) The place where new inmates are received, processed, tested, and assigned to housing.

certification Using a legal process to treat juveniles as adults for criminal prosecution.

certiorari To be informed of, to be made certain in regard to. The name of a writ of review or inquiry; a writ directed by a superior court to an inferior court asking that the record of a case be sent up for review; a method of obtaining a review of a case by the U.S. Supreme Court.

change of venue The removal of a action begun in one county or district to another county or district for trial.

charge To initiate formal criminal court proceedings; to impose a burden, duty, or obligation; to claim, demand, or accuse; to instruct a jury on matters of law.

charge bargaining Plea negotiations over the charge the government will file.

chronic stress The stress associated with the long-term effect of experiencing continual pressures and problems in the work environment.

circumstances The attendant facts. Any fact may be a circumstance with reference to another fact.

circumstantial evidence All evidence of an indirect nature. The existence of a principal fact is inferred from circumstances.

citation An order to appear in court; or, in writing, the citation of a documentable source.

citizen One who under the constitution and laws of the United States or of a particular state is a member of the political community.

civil liability Being held accountable in a civil court of law, where nominal, compensatory, and/or punitive damages can be awarded against those held liable for actions or in actions that resulted in harm, injury, or death.

civilian review boards Commissions set up outside the police or corrections department to hear and review citizen complaints against the police or the correctional system.

classification   The separation of inmates at all levels into groups according to characteristics that they share in common.

client   Someone who is under the care, custody, or control of a correctional agency.

closed-ended question   Questions in a questionnaire that force the respondent to select from a list of possible responses (also called *forced-choice questions*).

co-correctional institutions (also known as coed prisons)   Correctional facilities where both men and women are housed within one compound. Although they do not share living quarters in the United States, males and females interact socially and have access to the same institutional programs.

coercion   Compulsion, constraint, or compelling by force.

cognitive dissonance   A perceived discrepancy between what is stated to be reality and what is reality in fact.

colloquy   The formal discussion between judge and defendant to determine if defendants have pleaded knowingly and voluntarily.

commercial bail   The private business of bail bonding.

common law   A custom translated into law over time and social circumstances.

communication and feedback systems   The kind and amount of information flow among people, within an organization, and between an organization and people.

community policing   Citizen participation in setting police priorities and police operations.

community residential centers   Minimum-security community-based residential facilities that typically provide such programs as work release, drug treatment, and educational opportunity.

community service internships   Field experiences offering officers or students opportunities to gain broad perspectives on crime and criminal law enforcement.

community-based supervision   Services, programs, or facilities provided within the community to offenders who are not incarcerated in high-security confinement.

complaint   In criminal law, a charge preferred before a magistrate having jurisdiction that a person named has committed a specific offense. Usually, the first document filed with a court charging the offense. In some states, the term *complaint* is interchangeable with information; it is also sometimes used interchangeably with *affidavit*. *See also* information.

concept   A general idea in management stated in a formalized manner so that it can be communicated in a standardized fashion.

concurrent sentences   Sentences run at the same time and each day served by the prisoner is credited on each of the concurrent sentences.

conditional release   The release of an offender from incarceration under certain conditions, violation of which allows for reactivation of the unserved portion of the sentence.

conditioning   The expectation that a certain reaction will follow a certain stimulus, which is reinforced by repetition of the stimulus/response pattern.

conflict perspective   Conflict, not agreement, is the normal state of society. Crime is the product of the power structure.

conglomerate   A complex organization composed of numerous diverse functions.

conjugal visits   The authorization of visits that involve sexual intimacy.

consecutive sentences   Sentences that are served one after the other. Inmates refer to such sentences as *stacked* or as *boxcars*.

consensus perspective   General agreement on values in society.

consensus prison management   A balance of control and responsibility management.

consent decree   A response to a lawsuit whereby the court agrees to delay direct intervention in exchange for voluntary compliance with certain stipulated conditions.

consent search   A search conducted with knowing and voluntary consent.

consistent supervisory style   An approach to inmate management in which the correctional officer responds in a uniform manner whenever similar situations are encountered.

conspiracy   The agreement between or among parties to commit a crime.

constructive intent   The conscious or unconscious creation of risk of harm.

contact visits   The authorization of physical contact (within specified limits) during visits in jails and correctional centers.

contempt   A willful disregard or disobedience of a public authority.

contempt of court   Any act that is calculated to embarrass, hinder, or obstruct the court in the administration of justice, or which is calculated to lessen its authority or dignity. Directed contempts (also called *criminal contempts*) are those committed in the immediate view of the court (such as insulting language or acts of violence) and are punishable summarily. Constructive (or indirect) contempts are those which arise from matters not occurring in or near the presence of the court but with reference to the failure or refusal of a party to obey a lawful or decree of court.

contempt powers   The power of a court to punish for contempt. A court of record has this power.

contingency contracting   An agreement between parties (e.g., an inmate and the correctional administration) whereby one agrees to take specified action if the other meets certain conditions stipulated in the contract.

contraband   Any item that people are not authorized to possess, or an authorized item that is altered from its original state.

contract labor   The practice of using prisoners to work under contract to private industry.

control model of management   Prison management emphasizing obedience, work, and education of prisoners.

controlled movement   Restricting freedom of movement to better ensure institutional security.

controlling   Making certain that plans succeed by measuring and correcting activities of employees. Controlling is closely related to the organizational planning system.

conventional goals   Socially established goals (such as money, status, and prestige) that are recognized as desirable throughout society.

conviction   In a general sense, the result of a criminal trial that ends in a judgment or sentence that the person is guilty as charged.

corpus delicti   The body of the crime; the essential elements of the crime; the substantial fact that a crime has been committed; the actual commission by someone of the offense charged.

correctional boot camps   Shock incarceration for youthful first-time nonviolent offenders.

correlation   An association, but not necessarily causal.

count   The plaintiff's statement of his or her cause of action. Also used to specify the several parts of an indictment or information, each charging a distinct offense. Often used synonymously with the word *charge*.

counts   Periodic verifications of the total number of inmates in custody.

court above, court below   In appellate practice, the court above is the one to which a cause is removed for review, whether by appeal, writ of error, or certiorari; while the court below is the one from which the case is removed.

court of appeal   An appellate tribunal; the name given to the court of last resort in several states; the court of last resort of a particular type of case; or in some states, an intermediate appellate court below the supreme court.

court of common pleas   In English law, one of the four superior courts at Westminster. In U.S. law, the name given to a court of original and general jurisdiction for the trial of issues and law. The superior court of the District of Columbia is called the Court of Common Pleas.

court of competent jurisdiction   One having power and authority of law at the time of acting to do the particular act.

court of errors and appeals   The court of last resort in the state of New Jersey. Formerly, the same title was given to the highest court of appeal in New York.

court of general sessions   The name given in some states to a court of general original jurisdiction in criminal cases.

court of record   A court in which appeals are heard on the record. A court whose judicial acts and proceedings are recorded which has the power to fine or imprison for contempt.

court of special sessions   A court of inferior criminal jurisdiction in Oklahoma. Jurisdiction roughly equivalent to that of a justice of the peace.

court of star chamber   An English court of very ancient origin. Originally, its jurisdiction extended legally over riots, perjury, misbehavior of sheriffs, and other misdemeanors contrary to the laws of the land; afterward, stretched to the asserting of all orders of state; becoming both a court of law to determine civil rights and a court of revenue to enrich the treasury. It was finally abolished "to the general satisfaction of the whole nation."

court martial   A military court convened under the authority of the government and the Uniform Code of Military Justice for trying and punishing offenses committed by members of the armed forces.

courts of appeals A system of courts of the U.S. (one in each circuit) created by act of Congress, composed of three or more judges (provision also being made for allotment of the justices of the Supreme Court among the circuits) and having appellate jurisdiction as defined by statute. Called the U.S. Courts of Appeals; formerly called the Circuit Courts of Appeals or U.S. Circuit Courts of Appeals.

courts of the United States Comprise the Senate of the United States as a Court of Impeachment, the U.S. Supreme Court, the courts of appeals, the district courts, the court of claims, the court of customs and patent appeals, the customs court, the tax court, the provisional courts, and courts of territories and outlying possessions.

crime An act in violation of penal law. An offense against the state.

crime control The value or goal of reducing crime, emphasizing informal discretionary decision making.

crime control model Values discretion to quickly sort out factually innocent from factually guilty.

crimes cleared by arrest Crimes known to the police but removed from active police records.

criminal One who has committed a criminal offense. One who has been legally convicted of a crime. One adjudged guilty of a crime.

criminal action The whole or any part of the procedure that law provides for bringing offenders to justice.

criminal charge An accusation of crime in a written complaint, information, or indictment.

criminal court A court charged with the administration of the criminal laws and empowered to sentence the guilty person to fine or imprisonment. In New York, the criminal courts are comprised of the superior and local criminal courts. *Superior court* means the supreme court or a county court. *Local criminal court* means a district court, or the New York City criminal court, or a city court, or a town court, or a village court, or a supreme court justice sitting as a local criminal court or a county judge sitting as a local criminal court.

criminal event The commission of a specific crime.

criminal history A record of prior offenses.

criminal homicide All willful felonious homicides as distinguished from deaths caused by negligence.

criminal information A formal accusation of crime, differing from an indictment only in that it is preferred by a prosecuting officer instead of a grand jury.

criminal intent An intent to commit a crime; malice, as evidenced by a criminal act; an intent to deprive or defraud the true owner of his property.

criminal justice process The sequence of steps taken from the initial contact of an offender with the law until he or she is released back into a free society.

criminal justice system A loose confederation of agencies, including police, courts, and corrections.

criminal law That branch or division of law that deals with crime and punishment.

criminal negligence   The unconscious creation of a high risk of harm.

criminal procedure   The law prescribing how the government enforces criminal law. A method for the apprehension, trial, prosecution, and fixing the punishment of persons who have broken the law.

criminal proceeding   One instituted and conducted for the purpose of preventing the commission of crime, of fixing guilt for a crime already committed, and of punishing the offender.

criminal prosecution   An action or proceeding instituted in a proper court on behalf of the public for the purpose of securing the conviction and punishment of one accused of crime.

criminal recklessness   The conscious creation of a high risk of harm.

criminogenic forces   The causes of crime in the society, such numerous variables as poverty, socioeconomic status, or physical or mental impairment, among many others.

cross examination   The examination of a witness in a trial or hearing by the party opposed to the one who produced him, on the evidence given; to test its truth, to further develop it, or for other relevant purposes.

crowded prisoners   Prisoners who must live in less than 60 square feet of floor space.

curfew offenses   Offenses relating to violation of local curfew or loitering ordinances which provide regulations as to when a person (usually a juvenile) may lawfully be on the streets.

curtilage   The enclosed space of ground and buildings immediately surrounding the dwelling house.

custodial institution   A secure physical structure where offenders are confined with strict limitations on their access to free society.

cycle of violence   The hypothesis that childhood abuse creates a predisposition to later violent behavior.

D.A.R.E. (Drug Abuse Resistance Education)   Specially trained police officers are assigned to schools to teach drug prevention.

day in court   The opportunity to present one's claim before a competent tribunal.

daylight   That portion of time after sunrise and before sunset. Nighttime is the period between sunset and sunrise. Often important as to assessment of the degree of criminal culpability.

dealer   In the popular sense, one who buys to sell—not one who buys to keep.

decision-making process   Consists of who decides what is going to be done with the plan, how they decide, when and how fast they decide, and how their decisions will be put into action. Decision making also means who is going to solve problems and in what ways these problems are going to be solved.

decree   The judgment of the court; a declaration of the court announcing the legal consequences of the facts found.

decriminalization   Removing status offenses from juvenile jurisdiction.

defendant   The person defending or denying; the party against whom relief or recovery is sought in an action or suit. In criminal law, the party charged with a crime.

defense   In a criminal action, the answer made by the defendant to the state's case.

defense attorney   The attorney representing the accused in a criminal action.

defenses of excuse   To admit to the wrongfulness of crime but deny responsibility.

defenses of justification   To admit to the crime but assert that it was morally or ethically right to do it.

deferred release decisions   The setting of release after the determination that a prisoner has reformed and/or completed a task or judicial assignment.

deinstitutionalization   Community-based noninstitutional treatment as an alternative to incarceration.

delegation   The work a manager performs to entrust responsibility and authority to others and to create accountability for results. For example, a state director of corrections delegates authority and responsibility to a warden or area superintendent.

deliberate   As applied to a jury, the weighing of the evidence and the law for the purpose of determining the guilt or innocence of a defendant. In the case of jury sentencing, the deliberation may be for the purpose of fixing the sentence.

delinquency   Behavior that would be criminal if committed by adults.

delinquent juvenile   A person of no more than a specific age who has violated any law or ordinance or is incorrigible; a person who has been adjudicated a delinquent child by a juvenile court while of juvenile court age.

density   The number of square feet of floor space per prisoner.

deprivation model   An explanation of the prisonization process which maintains that it is a function of adapting to an abnormal environment that is characterized by numerous deprivations.

descriptive guidelines   Sentencing ranges based on actual past sentence practices.

detainer   A kind of "hold order" filed against an incarcerated man by another state or jurisdiction, which seeks to take the person into custody to answer to another criminal charge or conviction whenever he is released from the current imprisonment.

detention centers   Secure, temporary holding facilities, usually designated for juvenile offenders. A municipal jail.

diagnostic reception center   A central intake location where newly arriving inmates are interviewed, tested, examined, and evaluated for classification purposes.

differential association   Criminal behavior depends on association.

differential response   Police response to routine calls differs from that to emergency calls.

diminished capacity   Mental impairment less disabling than insanity.

direct evidence   That means of proof which tends to show the existence of a fact in question without the intervention of the proof of any other fact. Is distinguished from circumstantial evidence, which is often called *indirect*.

direct examination   In practice, the first interrogation or examination of a witness, on the merits, by the party on whose behalf he is called.

direct information   Facts known by direct knowledge.

direct supervision   Officers are in constant direct contact with the prisoners they supervise.

directing   Guiding, overseeing, coaching, and leading people toward goals and objectives while staying within the policies, procedures, and standards of the organization. Directing, more than any other management function, involves functional personal relationships.

discharge   The removal of a client from supervision, generally as a result of satisfactory completion of the conditions of probation or parole.

discretion   Decision making without formal recourse to laws and other written rules.

disorderly conduct   Conduct against public order. Sometimes used synonymously with *breach of peace*, although not all disorderly conduct is a breach of peace.

disposition hearing   To determine what treatment and custody should follow finding of delinquency.

district attorney   In many states, a district attorney or an assistant district attorney, and where appropriate, the attorney general or an assistant attorney general.

diversion   Removal from juvenile justice system to alternative programs, or transferring defendants into some alternative to criminal prosecution.

doubt   Uncertainty of mind; the absence of a settled opinion or conviction; the state of the case which, after the entire comparison and consideration of the evidence, leaves the minds of the jurors in such a condition that they cannot say with a moral certainty, of the truth of the charge. If upon proof there is a reasonable doubt remaining, the accused is entitled to the benefit of an acquittal.

dual system of justice   Separate systems for adults and juveniles.

duces tecum   From the Latin "bring with you." A subpoena *duces tecum* requires a party to appear in court and bring with him certain documents, pieces of evidence, or other matters to be inspected by the court.

due process   The value of formal rules and procedures to limit the power of government and protect the rights of individuals.

due process clause   The guarantee of fair procedures and protection of life, liberty, and property.

due process model   Emphasizes formal legal adversary process at the heart of the criminal process.

due process of law   The fundamental rights of the accused to a fair trial; the prescribed forms of conducting a criminal prosecution; the safeguards and protection of the law given to one accused of a crime. In substantive criminal law, the right to have crimes and punishments clearly defined in the law. Government can act only according to rules.

due process revolution   The expansion by the Supreme Court during the 1960s of the rights of criminal defendants and the application of the rights to state proceedings.

duress   To commit a crime under coercion.

d. w. i. (driving while intoxicated)   Driving or operating any motor vehicle while drunk or under the influence of liquor or narcotics.

effectiveness   The measurement of a program, plan, or effort in terms of its result or impact and not in terms of its resource cost. A program could, therefore, be highly effective (in terms of client service) but not efficient in terms of dollars, time, or other costs. Usually, programs are best measured in terms of both effectiveness and efficiency in order to attain a favorable benefit-cost ratio. However, this is not always possible in matters of custody or treatment.

efficiency   Planning in an organization and its work so that objectives can be attained with the lowest possible costs, which may mean money costs, human costs, or other resource costs.

embezzlement   The misappropriation or misapplication of money or property entrusted into one's care, custody, or control.

empiricism   The idea that all knowledge results from sense experience; a scientific method that relies on direct observation and the analysis of data.

equal protection of the law   Prevents unreasonable classifications.

et al.   And elsewhere; and others.

ethnographic study   Research by intensive field observation and interviews.

ethnography   The observational description of a people or some other social unit.

evaluation, impact   Showing whether the program, after implementation, has helped or not helped the group of people for whom the program was intended. Sometimes referred to as *product evaluation.*

evaluation, process   Assessment of a plan during the time it is being implemented. Process evaluation should be done at least weekly, and some critical parts of the plan should be assessed daily. This allows the planner to stay "on top" of the plan as it is being put into action.

evaluation research   Research to measure the effectiveness of a social program or institution.

evidence   Any species of proof, presented at the trial for the purpose of inducing belief in the minds of the court or jury.

ex parte   On one side only; by or for one party; done for; in behalf of, or on the application, of one party only.

ex post facto   After the fact.

ex post facto design   An after-only evaluation research design where pretesting is not possible.

ex post facto law   A law passed after the occurrence of a fact or commission of an act that retrospectively changes the legal consequences or relations of such fact or deed. A retroactive laws. Forbidden to both the states and the federal government by the U.S. Constitution.

ex rel   By or on the information of. Used in case title to designate the person at whose instance the government or public official is acting.

exception, management by   A feature of delegation where routine and frequently recurring matters should be handled by subordinates, allowing the manager to concentrate time and energies on exceptional and very important matters.

exclusionary rule   The rule that excludes from the trial of an accused, evidence illegally seized or obtained; prohibiting the use of illegally obtained evidence to prove guilt.

exclusive jurisdiction   The sole authority to hear and decide cases.

executive clemency or pardon   The authority of presidents and governors to eliminate a sentence.

existing statistics   Created statistical data that are available to researchers for analysis.

experiment   A research method that seeks to isolate the effects of an independent variable on a dependent variable under strictly controlled conditions.

experimental group   The group in an experiment that is exposed to the experimental treatment.

experimental mortality   Loss of subjects in an experiment over time. This is a potential cause of internal validity problems.

expert evidence   The testimony given in relation to some scientific, technical, or professional matter by experts (i.e. persons qualified to speak authoritatively by reason of their special training, skill, or familiarity with the subject).

expert witness   One who gives the results of a process of reasoning that can be mastered only by special scientists; one who has skilled experience or extensive knowledge in his calling or in any branch of learning; person competent to give expert testimony.

express bargaining   A direct meeting to decide concessions.

external validity   The generalizability of an experiment to other settings, other treatments, other subjects.

face validity   A form of content validity; a careful consideration and examination of the measurement instrument is made to determine whether the instrument is measuring what it purports to measure.

face-to-face interview   A method of administering a survey in which an interviewer questions an interviewee using a structured set of questions. *See also* interview schedule.

factorial design   The design of an experiment in which more than one independent variable is being measured.

factual guilt   Defendant has actually committed a crime, or has knowledge of guilt but not necessarily provable in court.

family crime   Crimes against people known to the offender.

federal question   A case that contains a major issue involving the U.S. Constitution or statutes. The jurisdiction of the federal courts is governed, in part, by the existence of a federal question.

federalism   The division of power between federal and state governments.

felonies   Serious crimes punishable by one year or more in prison. Crimes of a graver or more atrocious nature than those designated as a misdemeanor.

felony complaint   A verified written accusation by a person, filed with a local criminal court, which charges one or more defendants with the commission of one or more felonies and which serves to begin a criminal action but not as a basis for prosecution thereof.

field experiment   An experiment taking place in a real-world environment, where it is more difficult to impose controls.

field research   A research method based on careful observation of behavior in a natural social environment.

focus group   A small group of individuals drawn together to express views on a specific set of questions in a group environment. This method may serve a number of functions in social research; as a starting point for developing a survey, to recognize potential problems in a research design, or to interpret evidence.

follow-up research procedures   The methods of following up nonrespondents to mail questionnaires to increase response rate. Methods include sending postcard reminders, sending second questionnaires and request, and telephoning to solicit cooperation or to get the responses over the telephone.

forcible rape   Rape by force, or against the consent of the victim.

forecasting   The work a manager performs to estimate the future.

forgery   Making, altering, uttering (passing), or possessing anything false that is made to appear true, with intent to defraud.

formal criminal justice   The law and other written rules that determine the outer boundaries of action in criminal justice.

formalization   Replacing discretion with rules.

formative evaluation   An evaluation of a program in process, information from which will be used to reform or improve the program. *See* summative evaluation.

frequency distribution   The distribution of cases across the categories of a variable, presented in numbers and percentages.

frisk search   The physical pat-down of a clothed subject to determine whether weapons or other contraband items are concealed externally within clothes, shoes, hair, mouth, and so on.

fruits of a crime   Material objects acquired by means of and in consequence of the commission of a crime, and sometimes constituting the subject matter of the crime.

function   The total of positions encompassing one kind of work grouped to form an administrative unit. A group or family of related kinds of management work, made up of activities that are closely related to one another and have characteristics in common derived from the essential nature of the work done.

functional unit management   A decentralized approach wherein a unit manager, case manager, and counselor, along with supportive custodial, clerical, and treatment personnel, maintain full responsibility for providing services, making decisions, and addressing the needs of inmates assigned to a living unit.

fundamental fairness doctrine   A due process definition focusing on substantive due process.

funnel effect   The result of sorting decisions that lead to fewer individuals remaining at successive stages in the criminal justice process.

furlough   The privilege granted of temporary release from confinement, with the understanding that an inmate will return to the institution at a given time.

gambling Promoting, permitting, or engaging in gambling.

general deterrence To prevent crime in general population by threatening punishment.

general intent The intent to commit the actus reus.

general jurisdiction The authority to hear and decide all criminal cases.

general population inmate Prisoner without special problems.

general principles of criminal law The broad general rules that provide the basis for other rules.

goal The broadest, most long-range statement, in management terms, of the purpose or mission of an organization or unit within the organization. *See also* objective.

goal maintenance The process of working toward an established goal according to a plan, the application and guidance of resources toward the goal, and assessment of the degree and rate of progress toward attainment of the goal. It is appropriate within this process to redefine or redevelop goals as necessary.

goal setting The identification of individual or organizational purposes and intent; their specification as to time, resources needed, planning, and how to measure and report results.

good time Days deducted from prison terms based on good behavior of prisoners.

good-time laws The reduction of sentence length by one third or one half based on behavior in prison.

grand jury A jury of inquiry authorized to return indictments. Citizens who test the government's case and agree or disagree on prosecutable indictments.

grand larceny Larceny of the grade of felony, generally expressed in dollar value of amount stolen.

gross misdemeanors Crimes punishable by jail terms of 30 days to a year.

group home A relatively open, community-based facility.

guided discretion statutes Laws requiring juries to use guidelines on mitigating and aggravating circumstances.

habeas corpus (literally, "you have the body") *See* writ of habeas corpus.

halfway house Institutions in the community for parolees and probationers.

hands-off doctrine Prison management left to discretion of prison administrators.

Hawthorne principle The finding that creation of a new and closely watched project produces temporary positive results.

hearing In a broad sense, whatever takes place before a court or a magistrate clothed with judicial function and sitting without a jury. A trial is a hearing, but not all hearings require the formalities of a trial.

hearsay Information acquired through a third person; evidence offered by someone who does not know its truth firsthand.

home confinement A sentence to detention at home except for work, study, service, or treatment.

homicide The killing of one human being by another.

homogeneous groups   In sampling, strata formed by sets of individuals who share certain characteristics (gender, race, age, etc.).

hung jury   A jury so irreconcilably divided in opinion that it cannot agree upon any verdict.

hypothesis   A conditional statement relating the expected effect of one variable on another, subject to testing.

impeachment   A criminal procedure against a public officer to remove him or her from office. In the law of evidence, the adducing of proof that a witness is unworthy of belief.

importation hypothesis   The theory that prison society has its roots in the criminal and conventional societies outside the prison.

in re   In the affair; in the matter of; concerning. This is the usual method of entitling a judicial proceeding in which there are no adversary parties. For this reason, used in the title of cases in a juvenile court.

incapacitation   To prevent crime by incarceration, mutilation, or capital punishment.

incident report   A patrol or a correctional officer's description of a crime or broken regulation, usually detailing witnesses and suspects.

incident-based reporting   Reporting of each offense separately, whether part of the same event or not.

incident-driven strategies   Isolated event determines response by officers.

incorporation doctrine   Due process focusing on procedural regularity.

independent variable   The variable, in an experiment or survey, that exercises an effect on a dependent variable. The *cause* in a cause-and-effect model.

indeterminate sentence   An open-ended penalty tailored to the needs of individual offenders.

index   A composite measure developed to represent different components of a concept.

index crimes   The crimes used by the Federal Bureau of Investigation in reporting the incidence of crime in the U.S. in the Uniform Crime Reports. The statistics on the Index Crimes are taken as an index of the incidence of crime in the U.S.

indicators   Observable phenomena that can be used to measure dimensions of a concept.

indictment   The formal accusation of a crime by a grand jury.

indigenous theory   The belief that conditions inside a prison shape prison society.

indigent defendants   Defendants too poor to afford a lawyer.

inducement test   Entrapment focusing on government actions.

infamous crime   A crime that reflects infamy on the one who has committed it; crimes punishable by imprisonment in the state prison or penitentiary. At common law, all felonies were considered to be infamous crimes.

inferences   An accurate guess or conclusion based on evidence gathered on a relatively small probability sample, extrapolated to a much larger population.

inferential statistics   Statistics that allow a researcher to draw conclusions regarding the general population from the findings of a representative sample drawn from that population; statistics that utilize probability in decision making; hypothesis-testing statistics.

informa pauperis   In the form of a pauper; as a poor person or indigent. Permission to bring legal action without the payment of required fees for counsel, writs, transcripts and the like.

information   An accusation exhibited against a person for some criminal offense, without an indictment. An accusation in the nature of an indictment, from which it differs only in being presented by a competent public officer on his or her oath of office, instead of a grand jury on their oath. Formal accusation of a crime by a prosecutor.

informed consent   This is achieved when subjects in a research study comprehend its objectives, understand their level of confidentiality, and agree to cooperate.

informer   A person who informs or prefers an accusation against another whom he or she suspects of a violation of some penal statute.

infraction   The name given to minor offenses (chiefly traffic offenses) in the California Infractions Code.

initial case screening   Prosecutors reviewing whether to charge, divert, or dismiss a case.

injunction   A writ prohibiting an individual or organization from performing some specified action.

insanity   The legal term excusing criminal liability; not synonymous with mental illness.

institutional review board   Committees in institutions where scientific research is being carried out who review the research methods to be sure that the rights of human (or animal) subjects are being protected.

institutional support   Prisoners who work in maintaining the jail to pay part of the expenses of incarceration.

instrumentalities of a crime   The tools or implements used to commit a crime.

intake   The initial juvenile court process following a serious infraction of the law, unless preceded by diversion.

intensity structure   The patterns that make best sense of the multiple items in a scale, and their interrelation.

intensive probation supervision   Closely supervised probation, stressing retribution, incapacitation, and economy.

interaction effect   The tendency for a third variable to interact with the independent variable, thereby altering the relationship of the independent variable to the dependent variable. This means that the relationship between the independent and dependent variables will vary under different conditions of the third variable.

intermediate appellate courts   Courts that hear initial appeals.

intermediate punishment   Sanctions somewhere between the extremes of incarceration and straight probation.

intermittent incarceration   Incarceration at night and on weekends with release for school, work, treatment, or community service.

internal affairs unit   Units created to investigate, report, and recommend with respect to civilian complaints against police officers.

internal grievance mechanisms   Procedures inside prisons for dealing with grievances.

internal validity   The extent to which an experiment actually has caused what it appeared to cause.

interrogation   The process used in questioning suspects, usually after arrest and prior to filing charges.

intersubjectivity   The shared perceptions of individual observers. The greater the intersubjectivity, the greater the validity and reliability of the observations.

intervening variable   A third variable in a trivariate study that logically falls in a time sequence between the independent and dependent variables.

interview schedule   A set of questions with guided instructions for an interviewer to use in carrying out an interview.

invasion of privacy   A possible abuse in social research, in which rights of privacy have been ignored. Must be weighed in relation to the public's right to know. *See also* informed consent.

investigatory stage   In police practice the stage of investigation during which the offense is the subject of general inquiry before suspicion has focused on a particular person or persons. Distinguished from the *accusatory stage*, which covers the investigation that occurs after suspicion has focused on one or more particular individuals as being guilty of the offense.

issue   A single, certain, and material point, deduced from the pleadings of the parties, which is affirmed by one side and denied on the other; a fact put in controversy by the pleadings; in criminal law a fact that must be proved to convict the accused, or which is in controversy.

item analysis   A test for validity of an index in which a cross tabulation of total index scores to separate items making up the index is examined.

jail time   The credit allowed on a sentence for the time spent in jail awaiting trial or mandate on appeal.

judge   An officer so named in his commission, who presides in some court.

judgment   In general, the official and authentic decision of a court of justice upon the respective rights and claims of the parties to the action or suit therein litigated and submitted to its determination.

judicial process   The sequence of steps taken by the courts in deciding cases or disposing of legal controversies.

jurisdiction   The power conferred on a court to hear certain cases; the power of the police or judicial officer to act. The extent of the power of a public official to act by virtue of his or her authority.

jury panel   A list of jurors returned by a sheriff, to service at a particular court or for the trial of a particular case. The word may be used to denote either the entire body of the persons summoned as jurors for a particular term of court, or those selected by the clerk by lot.

justice model   Justice demands punishment for the crime committed. Focus on rights and rules in corrections.

juvenile delinquent   A youth who has committed either a status or a delinquency offense.

labeling theory   Society's response to crime defines some people as criminals.

laboratory experiment   An experiment taking place in a laboratory setting, where it is possible to maintain a large number of controls.

larceny   The taking of property from the possession of another with intent of the taker to convert it to his or her own use. Depending on the value of the property taken, the offensive is a felony or a misdemeanor.

law   Law is the formal means of social control that involves the use of rules that are interpreted, and are enforceable, by the courts of a political community. Law is the effort of society to protect persons, in their rights and relations, to guard them in their property, enforce their contracts, hold them to liabilities for their torts, and punish their crimes by means of sanctions administered by government.

leader   A person who enables other people to work together to attain identified ends.

leadership   The guidance and direction of the efforts of others. In management, the work of planning, organizing, directing, staffing, and controlling performed by a person in a leadership position to enable people to work most effectively together to attain identified ends.

leadership evolution   The systematic and continuing adaptation of a leader to the needs of the person, group, or organization.

leading questions   Questions that steer witnesses to a desired answer.

legal   Conforming to the law; according to a law; required or permitted by law; not forbidden or discountenanced by law; good and effectual law.

legal duty   That which the law requires to be done or forborne.

legal ethics   Usages and customs among the legal profession, involving their moral and professional duties toward one another, toward clients, and toward the courts.

legal guilt   Proof beyond a reasonable doubt by admissible evidence.

legal provocation   Provocation sufficient in law to be a defense to the act. Example: justifiable homicide.

legalistic style   Emphasis on criminal law enforcement and formal rules.

legislation   Rules of general application, enacted by a law-making body in a politically organized society. Included in legislation are constitutions, treaties, statutes, ordinances, administrative regulations, and court rules. Distinguished from case law, common law, and "judge-made law."

legislative sentencing model   Legislatures set penalties for offenses.

lesser included offense   A crime committed in the process of committing a crime of more serious degree or grade.

lesser offense   Sometimes used synonymously with a *less serious offense*, or *minor offense*.

levels of measurement   The four commonly defined levels for measuring variables: nominal, for distinct categories with no order; ordinal, for ordered categories; interval, for numerical scales with mathematically defined intervals between points on the scale but no true zero point; and ratio, for numerical scales with mathematically defined intervals and a true zero point.

limitation of actions   The time at the end of which no action at law can be maintained; in criminal law, the time after the commission of the offense within which the indictment must be presented or the information filed.

limited jurisdiction   Courts limited to hearing and deciding minor offenses and preliminary proceedings in felonies.

linear relationship   Shows that an increase (or decrease) in one variable is related to an increase (or decrease) in the other indicated by a diagonal best-fit line in a scattergram.

line-up (also called a *show up*)   A police identification procedure during which the person of a suspect is exhibited, along with others, to witnesses to the crime to determine whether or not they can connect him with the offense.

literature review   In a research project, the task of canvassing publications, usually professional journals, in order to find information about a specific topic.

local criminal courts   *See* criminal court.

local legal culture   The attitudes, values, and expectations toward law and legal practice in specific communities.

lockdown   The suspension of all activities, with prisoners confined to their cells.

longitudinal data   Data gathered over time.

longitudinal designs   Studies based on longitudinal data include trend studies, in which data are compared across time points on different subjects; cohort studies, in which data on subjects from the same age cohort are compared at different points in time; and panel studies, in which the same subjects are compared across time points.

mail survey   A survey consisting of a self-administered questionnaire, instructions, and a request for participation sent out through the mail to a selected sample.

mala in se   Acts or crimes immoral or wrong in and of themselves.

mala prohibita   Crimes wrong because a statute defines them as wrong, although no moral turpitude may be attached, and constituting crimes only because they are prohibited.

management development   The work a manager performs to help managers and candidates for management positions to improve their knowledge, attitudes, and skills.

manager   The person in the organization who may be responsible for any or all of the following: (1) the outcome of her job, (2) the outcomes of some other people's jobs (subordinates), (3) some of the outcomes of other people's jobs (peers, staff, and other managers), and (4) outcomes of activities of some persons outside the organization. A *professional manager* is one who specializes in the work of planning, organizing, directing, staffing, and controlling the efforts of others and does so through systematic use of classified knowledge, and a common vocabulary and principles, and who subscribes to the standards of practice and code of ethics established by a recognized body.

mandatory minimum sentence legislation   The requirement that judges must sentence offenders to a minimum time in prison.

mandatory parole release statutes   Laws requiring the release of prisoners at specified times.

mandatory release   Release based on good behavior and other sentence-reducing devices.

manslaughter   The lowest degree of culpable homicide death caused by culpable recklessness or negligence.

matching   An experimental procedure in which subjects to be placed in the experimental group are matched with subjects possessing similar characteristics in the control group.

material allegation   An allegation essential to the claim of defense, which could not be stricken from the pleading without leaving it insufficient.

material fact   A fact that is essential to a case, defense, or application, without which it could not be supported.

matrix questions   Sets of questions in a questionnaire that use the same set of response categories.

maximum-security prisons   Prisons that focus on preventing prisoners from escaping or hurting themselves or others.

measured capacity   One prisoner per cell.

measurement   A process in which numbers are assigned according to rules of correspondence between definitions and observations.

measurement error   Error unavoidably introduced into measurement in the process of observing a phenomenon. An observed measure (or score) is therefore based on the true score plus or minus the error. In social research this error may necessarily be great because of the crudity of the instruments used in measuring social phenomena.

medical model   Views crimes as a disease that requires treatment to cure.

medium-security prisons   Prisons that focus less on security and allow prisoners greater freedom of movement.

mens rea   A guilty mind; a guilty or wrongful purpose; a criminal intent. Guilty knowledge and willfulness.

merit system   The selection of judges by a governor from a list drawn up by a commission of citizens, lawyers, and judges.

middle-range offenders   Those not requiring imprisonment but demanding more than ordinary probation.

minimum-security prisons   Prisons containing prisoners who do not pose security problems and can therefore emphasize trust and a normal life-style.

minor   A person or infant who is under the age of legal competence; one under 21.

Miranda warning   The warning that must be given to a suspect whenever suspicion focuses on him. The officer must warn the suspect (1) that he or she has the right to remain silent; (2) that if the suspect talks, anything that he or she says may be used against him or her; (3) that he or she has the right to be represented by counsel and the right to have counsel present at all questioning; and (4) that if he or she is too poor to afford counsel, counsel will be provided at state expense.

misdemeanor   Any offense that is not a felony, punishable by one year or less in jail.

misdemeanor complaint   As defined in several states, a verified written accusation by a person, filed with a local criminal court, which charges one or more defendants with the commission of one or more offenses, at least one of which is a misdemeanor and none of which is a felony, and which serves to begin a criminal action but which may not, except upon the defendant's consent, serve as a basis for prosecution of the offenses charged therein.

mistake of fact   Ignorance or error concerning facts.

mistake of law   Ignorance or mistake concerning the law.

moot   A subject for argument; unsettled; undecided. A moot point is one not settled by judicial decision.

moot case   A case that seeks to get a judgment on a pretended controversy, or a decision in advance about a right before it has actually been asserted and contested, or a judgment on some matter which, when rendered, for any reason, cannot have any practical legal effect on a then-existing controversy.

moot court   A court held for the arguing of moot (or pretended) cases or questions such as by students in law school.

moral turpitude   An act of baseness, vileness, or depravity in the private and social duties that man owes to his fellow man, or to society in general, contrary to the accepted and customary rule of right and duty between man and man.

motivating   The work a manager performs to inspire, encourage, and impel people to take desired action.

murder   The highest degree of culpable homicide.

narcotic offenses   Offenses relating to narcotic drugs, such as unlawful possession, sale or use. Also used to describe any substance abuse offense.

narcotics   Drugs, such as morphine or heroin, that in medicinal doses relieve pain and induce sleep, and in toxic doses cause convulsions, coma, or death.

National Crime Victim Survey (NCVS)   A national sample of victims surveyed about their victimization.

National Institute of Justice   The research arm of the U.S. Department of Justice.

natural experiment   An experiment that has not been brought about by the efforts of the experimenter but has occurred naturally in the real world and is being selected out for study by the experimenter.

negative evidence   In a field study, the nonoccurrence of expected events, an occurrence that is not reacted to, or one that is distorted in its interpretation or withheld from analysis.

negative (inverse) relationship   A type of relationship between two variables in which cases that are low on one variable are high on the other. *See also* positive (direct) relationships.

negotiated plea   A plea of guilt in exchange for a concession by the government.

negotiation   A give-and-take activity that allows individuals or groups to agree in common to a set of objectives, tasks, and shared use of resources.

net widening   Expanding jurisdiction, such as when sentencing borderline cases to intermediate punishments instead of straight probation.

new-generation jail   A jail that combines architecture, management, and training to provide safe, humane confinement.

new-generation prisons   A prison that combines management and architecture to provide safe, secure confinement for maximum security prisoners.

nonequilvalent control group   A control group that was not selected on the basis of random assignment. Usually created as a rough comparison group to participants in a social intervention program under evaluation. *See also* ex post facto design.

null hypothesis   A logical assumption that there is no relationship between the two variables being studied in the population. This assumption can be tested with inferential statistics.

objective   A specific, time-framed, behavioral expression of some end result or end product that is reasonably attainable, yet sufficiently challenging. Objectives can be long range (several years) or very short range (a day or less).

occupancy   The number of prisoners for each unit of confinement as set by federal and state statutes.

occupational crime   Crimes committed in the course of employment.

operating work   The work a manager performs other than the planning, organizing, directing, staffing, and controlling work that logically belongs to that position.

opportunity theory   The belief that criminal behavior depends on the available criminal opportunities; noncriminal behavior, on noncriminal opportunity structure.

order or recognizance or bail   A securing order releasing a principal on his or her own recognizance or fixing of bail.

organization   Any group of people formally associated to plan, implement, or evaluate a program or idea. The "organization" can be a group of people in need of a program, an agency or office to meet a need, or any other group formed to help meet a need.

organization chart   A schematic representation of organization structure, authority, and relationships.

organization crime   Crimes committed to benefit organizations illegally.

organization structure   The pattern work assumes as it is identified and grouped to be performed by people.

organizing   Establishing a system for performance toward stated goals and objectives. Putting the organization into desired structure and order.

original jurisdiction   The authority to initiate proceedings. Jurisdiction in the first instance; jurisdiction to take cognizance of a case at its inception, impanel a jury, try the case, and pass judgment on the law and facts. Distinguished from *appellate jurisdiction.*

pardon An act of grace, proceeding from the power entrusted with the execution of the laws, which exempts the person on whom it is bestowed from the punishment the law inflicts for the crime committed.

parens patriae (literally, "father of his country") The doctrine that the juvenile court treats the child as "a kind of loving father." Government acts as parent.

parole A conditional release from prison. The release of a prisoner from imprisonment but not from legal custody of the state, for rehabilitation outside prison walls under such conditions and provisions for disciplinary supervision as the parole board or its agents may determine. Parole is an administrative act and follows incarceration.

parole board A panel of civilians and experts that determines the release from prison to parole.

particularity The detailed description in a warrant of the object of a search.

pendulum swing The alternating emphasis on crime control and due process in the history of criminal justice.

per curiam By the court. An opinion of the court that is authored by the justices collectively.

per se By himself or itself; taken alone.

performance appraisal A formal program comparing employees' actual performance with expected performance.

persons arrested A wide variety of serious and minor offenses reported in raw numbers.

petit jury A trial jury as distinguished from a grand jury; an ordinary jury of 12 men (or fewer) for the trial of a civil or criminal action.

petit larceny Larceny of the grade of misdemeanor.

petty misdemeanors Crimes punishable by fine or up to 30 days in jail.

plain view search Object of seizure discovered inadvertently where an officer has a right to be.

plan In management, a predetermined course of action.

planning Selecting from alternative courses of future action. Determination of goals to be accomplished and how and when they are to be achieved.

plea of guilty A confession of guilt in open court.

plea of nolo contendere (literally, "no contest") A plea of neither guilty nor not guilty of a charge in criminal court. One that has the same effect in a criminal action as a plea of guilty but does not bind the defendant in a civil suit for the same wrong.

plea of not guilty A plea denying the guilt of the accused for the offense charged and putting the state to the proof of all the material elements of the offense.

podular design Allows greater security and opportunity for surveillance of fewer numbers of prisoners.

police academy A training school where police socialization begins.

police corruption   A form of occupational crime in which officers use their authority for private gain.

police defensiveness   The distrust of outsiders, who may not understand the law enforcement policies and procedures.

police depersonalization   Treating violence and other unpleasant experiences as matter of fact.

police misconduct   A range of illegal behavior, including brutality, constitutional violations, corruption, and unfair treatment of citizens.

police stress   The negative pressures associated with police work.

police working personality   The character traits of police officers revealed in their work as usually identified by sociologists and psychologists.

police-prosecutor teams   Police officers and prosecutors working together from investigation to conviction.

policy   A standing decision made to apply to repetitive questions and problems of significance to an organization as a whole.

political community   A political community involves forcible maintenance of orderly dominion over a territory and its inhabitants.

population   The collection of all elements (either known or unknown) from which a sample is draw. In a probability sample, the population consists of the elements in the sampling frame.

position   Work grouped for performance by one person.

positive (direct) relationship   A type of relationship between two variables in which cases that are high on one variable tend to be high on the other, and cases that are low on one variable tend to be low on the other. *See also* negative (inverse) relationship.

positivist   A person who strives to accumulate facts as the sole means of establishing explanations.

posttraumatic stress syndrome   Mental impairment caused by stress during battle or some traumatic event.

precedent   An adjudged case or decision of a court of justice considered as furnishing an example of authority for an identical or similar case afterward arising on a similar question of law. *See also* stare decisis.

precoded questionnaire   Coding information that is included on the questionnaire instrument itself. This facilitates transferring the data to a computer.

preemptory challenge   Self-determined, arbitrary, requiring no cause to be shown. As applied to selection of jurors, challenges allowed by law to both the state and defense to remove a prospective juror without cause from the panel jurors.

preliminary hearing   The examination of a person charged with a crime before a magistrate.

preliminary jurisdiction   A criminal court has "preliminary jurisdiction" of an offense when regardless of whether it has trial jurisdiction thereof, a criminal action for such an offense may be begun therein, and when such court may conduct proceedings with respect thereto which lead or may lead to prosecution and final disposition of the action in a court having trial jurisdiction thereof.

preponderance of the evidence   Greater weight of evidence. The preponderance of the evidence rests with the evidence that produces the stronger impression and is more convincing as to its truth when weighed against the evidence in opposition.

prescriptive guidelines   Sentencing ranges prescribing new practices.

presentment   The initial appearance by the accused before the magistrate after arrest. Also, a written notice taken by a grand jury of any offense, from their own knowledge or observation, without any bill of indictment laid before them at the suit of the government. *See also* indictment.

presumption of fact   An inference affirmative or disaffirmative of the truth or falsehood of any proposition or fact. Presumptions of fact are not the subject of fixed rules but are merely natural presumptions such as appear from common experience to arise from the particular circumstances.

presumption of innocence   To treat all individuals as innocent until proven guilty according to legally correct proceedings.

presumption of law   A rule of law that courts and judges shall draw a particular inference from a particular fact, or from particular evidence, unless and until the truth of such inference is disproved; and inference that the court will draw from the proof, which no evidence, however strong, will be permitted to overcome. Presumptions of law are reduced to fixed rules and from a part of the system of jurisprudence to which they belong. Presumptions are evidence, or have the effect of evidence.

preventive detention   The detention of defendants prior to trial to protect public safety.

preventive patrol   Moving through the streets to intercept and prevent crime.

prima facie case   A case developed with evidence such as will suffice until contradicted and overcome by other evidence.

prima facie evidence   Evidence good and sufficient on its face; such evidence as, in the judgment of the law, is sufficient to establish a given fact, or the group or chain of facts constituting the party's claim or defense, and which if not rebutted or contradicted will remain sufficient.

principle   A fundamental truth that will tend to apply in new situations in much the same way as it has applied in situations already observed.

principle of least eligibility   Prisoners should earn less than free citizens doing the same work, and should be less eligible than schoolchildren and welfare clients when competing for the same tax dollars.

prisoners' rights   Constitutional rights that survive incarceration.

privatization   The private management of correctional facilities.

pro bono assistance   The representation of criminal defendants without a fee.

pro se filings   Court proceedings in which prisoners file their own papers.

proactive police operations   Operations initiated by police.

probable cause   Reasonable cause; having more evidence for than against. An apparent state of facts that would induce a reasonably intelligent and prudent person to believe, in a criminal case, that the accused person had committed the crime charged.

More than suspicion, less than certainty. The quantum of proof required to search or arrest.

probation   The release of a convicted defendant by a court under conditions imposed by the court for a specified period during which the imposition of sentence is suspended. Probation is in lieu of incarceration and is a judicial act.

problem solving   The process of identifying and removing barriers to the setting and attainment of goals.

procedural due process   Limits on criminal procedure.

procedural law   The machinery for carrying on a suit or action.

procedure   A standardized method of performing specified work. The mode of proceeding by which a legal right is enforced as distinguished from the law, which gives or defines the right; the machinery, as distinguished from its product. A form, manner, and order of conducting prosecutions.

proof beyond a reasonable doubt   Enough facts to convict a criminal defendant.

prosecutor   One who prosecutes another for a crime in the name of the government.

prosecutor's information   As defined in New York, a written accusation by a district attorney filed with a local court, which charges one or more defendants with the commission of one or more offenses, none of which is felony, and which serves as a basis for prosecution thereof.

prostitution   Sex offenses of a commercialized nature.

protective custody units   Units devoted to the protection of prisoners with special problems.

provocation   The act of inciting another to do a particular deed; that which arouses moves, calls forth, causes, or occasions.

proximate cause   That which, in a natural and continuous sequence, is unbroken by any efficient intervening cause, produces the injury, and without which the result would not have occurred.

public defender   An attorney designated by law or appointed by the court to represent indigent defendants in criminal proceedings. A public defender is paid by the state or by private agency, or serves without fee.

public order offenses   Minor crimes of public annoyance.

public works crew   Prisoners who work in groups performing public services.

quality arrests   Arrests resulting in conviction.

quantification   Determining or measuring quantity or amount.

quantum of proof   The amount of evidence that justifies government action.

quasi-judicial proceedings   Proceedings that mix formal rules and discretionary judgments.

quasi-military lines   A form of bureaucracy with a hierarchical authority structure.

random sample   A sample in which any person or item in the population has an equal chance of being chosen on each selection.

rape   The unlawful carnal knowledge of a woman by a man forcibly and against her will.

rational decision making   Decisions based on defined goals, alternatives, and information.

real evidence   Evidence furnished by things themselves on view or inspection, as distinguished from a description of them given by a witness.

reasonable suspicion   The quantum of proof required for a stop-and-frisk procedure.

rebuttable presumption   A presumption that may be rebutted by evidence; a species of legal presumption that holds good until disproved.

rebuttal   The introduction of rebutting evidence; showing that a statement of witnesses as to what occurred is not true; the stage of a trial at which such evidence may be introduced; also the rebutting evidence itself.

rebutting evidence   Evidence given to explain, repel, counteract, or disprove facts given in evidence by the adverse party.

receiving stolen property   Buying, receiving, and possessing stolen property with knowledge that it is stolen or under circumstances requiring inquiry as to its origins.

recidivist   A repeat offender.

record   A written account of an act, transaction, or instrument, a written memorial of all the acts and proceedings in an action or suit, in a court of record; the official and authentic history of the cause, consisting in entries in each successive step in the proceedings. At common law, a roll or parchment on which the proceedings and transactions of a court are entered.

rehabilitation   To prevent crime by changing the behavior of individual offenders.

relative deprivation   Feelings of deprivation when compared to persons who are doing better.

release on (own) recognizance   The release of defendants on their promise to appear.

relevant   Applying to the matter in question. A fact is relevant to another fact when according to a common course of events, the existence of one taken alone or in connection with the other fact renders the existence of the other certain or more probable.

relevant evidence   Evidence that relates to the elements of a crime.

reported capacity   The number of prisoners that a jurisdiction decides is the capacity of a facility.

res gestae   Things done. The whole of the transaction under investigation and every part of it. *Res gestae* is considered an exception to the hearsay rule and is extended to include not only declarations by the parties to the suit but also statements made by bystanders and strangers under certain circumstances.

respondent   The defendant on appeal; the party who contends against an appeal.

response time   The time it takes for the police to respond to citizen calls.

responsibility   The responsibility of a subordinate to a superior for authority received by delegation; it is both absolute and tenuous. It is absolute as long as the subordinate maintains and executes the responsibility and authority appropriately. It is tenuous in that it can be taken back at any time by a higher authority in the organization. In any case, no superior can escape ultimate responsibility for any delegation or any activities of subordinates.

restitution   Repayment by offenders for the injuries their crimes caused.

retribution   Looks back in order to punish for the crime committed.

reus   A person judicially accused of a crime; a person criminally proceeded against.

revocation   The retraction of parole.

right of allocution   The right of the convicted person to speak in his own defense before judgment is pronounced. *See also* allocution.

right-wrong test   An insanity definition focusing on impairment of reason.

robbery   Stealing or taking anything of value from a person by force or violence or by putting in fear.

role   The specific relationship you have to other people or to an organization. Any given person can have many roles, depending on how many relationships they have or how many "hats" they wear in the organization. Role also has to be defined as a "two-way street"; half the role is how you see yourself in the relationship, and the other half comes from how the other person (or organization) sees you in the relationship.

role management   The continuous examination and assessment of roles. This could be one's own role, or it could be the roles of others as they affect your own role. The purpose is to modify or adjust your own role, or help others adjust their role, to maintain common interpersonal and organizational objectives.

rule of law   The principle that rules, rather than discretion, govern decisions in criminal law and procedure.

runaway   A juvenile offense; also an offender who has run away from home without his parents' permission.

safety-valve policy   Reducing the minimum sentence of prisoners when prisons exceed capacity.

sample   A small group that, ideally, is representative of a larger group.

sampling frame   That specific part of a population from which a sample is drawn for a survey.

scienter   Knowingly; with guilty knowledge.

Scottsboro case   The case that established the fundamental fairness doctrine.

search   Examining person or property to discover evidence, weapon, or contraband.

search incident to arrest   A search without a warrant conducted at the time of arrest.

Section 1983 actions   Legal actions brought under the Ku Klux Klan Act, permitting citizens to sue government officials for the violation of civil rights.

securing order   In New York, an order of a court committing a principal to the custody of the sheriff, or fixing bail, or releasing the person on his or her own recognizance.

selective hypothesis fallacy   Choosing subjects for research that favor a particular outcome.

selective incapacitation   The policy of imprisoning offenders who commit the most crimes.

self-reports   Collecting data by sampling members of the population who have committed crimes.

sentence   The judgment formally pronounced by the court or judge on a defendant after his or her conviction in a criminal prosecution, and stipulating the punishment to be inflicted.

sentence bargaining   Plea negotiations over the sentence a judge will grant.

sentencing discrimination   The determination of sentences by unacceptable criteria, such as race.

sentencing disparity   A difference in the sentences received by persons who committed similar offenses under similar circumstances.

sentencing guidelines   A range within which judges prescribe specific sentences.

separation of powers   The doctrine that permits the three branches of government—legislative, executive, and judiciary—to perform their own functions without interference from the others.

sequester   To keep a jury together and in isolation from other persons under charge of the bailiff while a trial is pending, sometimes called *separation of the jury*. To keep witnesses apart from other witnesses and unable to hear their testimony.

service of process   The service of writs, summonses, rules, and so on, signifies delivering or leaving them with the party to whom or with whom they ought to be delivered or left, and when they are so delivered, they are then said to be served.

sex offenses   Rape, prostitution, commercialized vice, statutory rape, and offenses against chastity, common decency, and morals.

shelters   Temporary, nonsecure, community-based holding facilities.

show cause   An order to appear as directed and to present to the court reasons and considerations as to why certain circumstances should be continued, permitted, or prohibited, as the case may be.

simple assault   Assault that is not of an aggravated nature. *See also* assault.

simplified traffic information   A written accusation by a police officer filed with a local criminal court which charges a person with a traffic violation or misdemeanors relating to traffic, and which may serve both to begin a criminal action for such offense and as a basis for prosecution thereof.

skill   A person's mental, emotional, and motor capacity to perform a certain function or task, which could range from the physical skill of operating an office machine to the emotional skills of working with people to the mental skill of computing a complex budget.

social control   The process by which subgroups and persons are influenced to conduct themselves in conformity to group expectations.

social control perspective   The view that obedience to rules depends on institutions to keep the desire to break the rules in check.

social structure of the case   Extralegal or sociological influences on decisions.

solicitation   Asking another person to commit a crime.

solvability factors   Information that leads to the solution of crimes.

special management inmates   Prisoners in need of special care.

specialization   The attempt to confine the work of each person to a single related set of functions, with sets of similar functions grouped together under one department or unit.

specific intent   The intent to do something in addition to the criminal act.

split sentence   Part of a sentence served in jail, the remainder served on probation.

split-sentence probation   A sentence to a specified term of incarceration followed by a specified time on probation.

staffing   Putting people into the proper jobs; the idea of having the right person in the right job at the right time. Staffing includes the selection, placement, development, and appraisal of people for organizational activities.

standing   The qualifications needed to bring legal action.

stare decisis   To abide by, or adhere to, decided cases; doctrine that when a court has once laid down a principle, it be applied to all future cases where facts are substantially the same, regardless of whether the person and the property are the same.

state   The supreme political community; also a state of the United States.

statistics   Figures that summarize and represent factual data.

status offenses   Behavior that only juveniles commit.

statutory law   All laws enacted by federal, state, or local legislatures.

statutory rape   Carnal knowledge of a female child below the age fixed by statute. Neither force nor lack of content are necessary elements of this offense.

stop and frisk   Less intrusive seizures and searches protected by the Fourth Amendment.

straight plea   A plea of guilty without plea negotiations.

strain theory   A belief that pressures in the social structure cause crime.

street crimes   One-on-one crimes against strangers.

strict liability   Criminal liability without criminal intent.

subculture of competition   The concept that success is more important than the means by which it is achieved.

subculture of violence   A subculture that condones violence.

subpoena   A process issued by a court to cause a witness to appear and give testimony for the party named.

substantial capacity test   An insanity definition focusing on impairments of either or both reason and will.

substantive due process   Constitutional limits on criminal law.

summons   A notification of proceedings against defendants and requirements of their appearance in court.

superior courts   Used generally to denote courts of general trial jurisdiction. The name given to felony courts in California and Illinois.

supervision   The day-to-day direct management of personnel and activities within a program. Each person and each activity should have an immediate supervisor who is responsible for the proper application of that person's skills and the proper direction of that activity.

supreme court   The highest court of the United States, created by the Constitution; the name given in most states to the highest court of appeals, the court of last resort.

suspect   To have a slight or even vague idea concerning; not necessarily involving knowledge of belief of likelihood; sometimes used in place of the word *believe*. Also, a person who is suspected of having committed an offense or who is believed to have committed an offense.

systems paradigm   The decision-making perspective that treats the criminal justice agencies as an integrated whole.

tasks   The specific items of activity for which each person in an organization is held accountable.

testimony   Evidence given by a competent witness, under oath or affirmation; as distinguished from evidence derived from writings and other sources. Testimony is one species of evidence, but the words *testimony* and *evidence* are often used interchangeably.

the great writ   A name given to the writ of habeas corpus.

theft   A popular name for larceny.

theory X   A management theory that employees can be motivated only by fear (of job loss, for example).

theory Y   A management theory that employees can be motivated by better challenges, personal growth, and improved work performance and productivity.

theory Z   A management theory that employees should be involved, should participate, and should be treated like family.

tort   A noncriminal legal wrong. A private or civil wrong or injury; a legal wrong committed upon a person or property independent of contract which is redressed in a civil court. A *personal tort* involves or consists of an injury to the person or to the reputation of feelings as distinguished from an injury or damage to real or personal property, called a *property tort.*

tort reasor   One who commits a tort.

training   The provision of a variety of ongoing opportunities for staff development, including coaching, workshops, seminars, and classes in higher education.

training schools   Secure detention facilities.

transcript of record   The printed record as made up in case for review by a higher court; also a copy of any kind. In referring to the written documents on appeal, the words *transcript, record,* and *record on appeal* are used interchangeably.

transferred intent   The concept that the intent to cause one harm results in causing harm to another.

trial jurisdiction   Jurisdiction by a criminal court of an offense when an indictment or an information charging such offense may properly be filed with such court, and when such court has authority to accept a plea to, try, or otherwise finally dispose of such as accusatory instrument. *See also* original jurisdiction.

typology   A classification of phenomena according to differing characteristics.

Uniform Crime Reports   A summary of information provided by local police agencies to the FBI.

unity of command   The principle that the more complete a reporting relationship a person has to a single superior, the less the problem of conflict in instructions and evaluation and the greater the feeling of personal responsibility for results.

unity of objectives   The concept that if persons in each position fulfill clearly defined objectives logically related to each other, the goal of the entire organization will be met.

utilitarian punishment   Looks forward to preventing crime in the future.

validity   The characteristic that a measuring instrument such as a survey has when it actually measures what it purports to measure.

values   The beliefs, attitudes, and expressed behavior that a person holds in terms of what they will accept, reject, or feel neutral toward. The body of history, policies, goals, leadership, and so on, that determine what an organization will produce as goods or services and how it will go about achieving that production.

vandalism   Willful or malicious destruction, injury, disfigurement, or defacement of property without consent of the owner or person having custody or control.

variables   The elements of an equation, experiment, or formula that are under study and subject to change in accordance with changes in the environment; anything that varies.

vehicle search   The search of vehicles without a warrant but not without probable cause.

venire (from the Latin for "to come," "to appear")   The name given to the writ for summoning a jury, and also the body of jury summoned.

venireman   A member of a jury; a juror summoned by the writ of venire facias.

venue   A neighborhood, place, or county in which an injury is declared to have been done or a fact declared to have happened. *Jurisdiction* of the court is the inherent power to decide a case, whereas *venue* designates the particular county or city in which a court with jurisdiction may hear and determine the case.

verdict   A formal and unanimous decision or finding made by a jury, impaneled and sworn for trial of a cause, and reported to the court upon the matters or questions duly summitted to them upon the trial. From the Latin *verdictum*, a "true declaration."

victimless crimes   Crimes without complaining victims.

violation   An incident punishable by a small fine that does not carry with it a criminal record.

violence   Physical force.

violent predators   Career criminals who commit a range of street crimes.

void for vagueness   Statutes must define crimes precisely.

voir dire (literally, "to speak the truth")   The preliminary examination of a witness or juror as to his competency, interest, and so on.

waive   To abandon or throw away; in modern law, to abandon, throw away, renounce, repudiate, or surrender a claim, privilege, or right, or the opportunity to take advantage of some defect irregularity or wrong.

warrant   A document issued by a magistrate that the Constitution requires for a search or arrest.

warrant of arrest   A written order issued and signed by a magistrate, directed to a peace officer or some other person specially named, commanding him or her to arrest the body of a person named in it, who is accused of an offense.

watchman style of policing   Focus on order maintenance and discretionary decision making.

work release   A program allowing prisoners to leave confinement to work.

writ of habeas corpus   A writ directed to a person detaining another and commanding him or her to produce the body of the prisoner or person detained.

zero-based budgeting   A method of budgeting that starts with no base from the preceding budget period. Most criminal justice vendors are subject to this method.

References

American Psychological Association. 1994. *Publication Manual of the American Psychological Association.* 4th ed. Washington, D.C.: American Psychological Association.

American Sociological Association. 1997. *American Sociological Association Style Guide.* 2d ed. Washington, D.C.: American Sociological Association.

American Sociological Review. 1936. Annual Washington, D.C.: American Sociological Association.

Bear, John and Mariah Bear. 2001. *Bear's Guide to Computer Degrees by Distance Learning.* New York: Ten Speed Press.

Becker, Howard S., Blanche Geer, Everett C. Hughes and Anselm L. Strauss. 1961. *Boys in White: Student Culture in Medical School.* Chicago, IL: University of Chicago Press.

Becker, Ronald F. 1997. *Specific Evidence and Expert Testimony Handbook: A Guide for Lawyers, Criminal Investigators and Forensic Specialists.* Springfield, IL: Charles C. Thomas.

Bouma, Gary D. and G. B. J. Atkinson. 1995. *A Handbook of Social Science Research: A Comprehensve and Practical Guide for Students.* 2d ed. New York: Oxford University Press.

Bray, R., ed. 1996. *Guide to Reference Books.* Chicago: American Library Association.

Brundage, D., R. Keane and R. Mackneson. 1993. "Application of Learning Theory to the Instruction of Adults." Pp. 131–144 in *The Craft of Teaching Adults,* edited by Thelma Barer-Stein and James A. Draper. Toronto, ON: Culture Concepts.

Carr, Sarah. 2000. "Online Psychology Instruction Is Effective but Not Satisfying, Study Finds." *Chronicle of Higher Education,* March 10, pp. 8–12.

Chicago Manual of Style. 1993. 14th ed. Chicago, IL: University of Chicago Press.

Commager, Henry S., ed. 1963. *Documents of American History.* 7th ed. New York: Appleton-Century-Crofts.

Doyle, Michael and Robert Meadows. 1997. "A Writing-Intensive Approach to Criminal Justice Education: The California Lutheran University Model." *Justice Professional* 10(1):19–30.

Edwards, Richard H., ed. 1995. *Encyclopedia of Social Work.* New York: National Association of Social Workers.

Hara, Noriko and Rob King. 1999. "Students' Frustrations with a Web-Based Distance Education Course." *First Monday: Peer Reviewed Journal on the Internet* 4(12):7–10.

Hartwell, Patrick. 1985. "Grammar, Grammars, and the Teaching of Grammar." *College English* 47:111.

Hunter, David E. and Phillip Whitten, eds. 1976. *Encyclopedia of Anthropology.* New York: Harper and Row.

Isaac, Stephen and William B. Michael. 1981. *Handbook in Research and Evaluation.* 2d ed. San Diego, CA: EdITS Publishers.

Johnson, Jr., William A., Richard P. Rettig, and Gary A. Steward. 1993. *Drugs, Self-Concept and Street Gangs as They Relate to Juvenile Delinquency: Phase II of a Drug Assessment Study with Young Offenders in a Secure Detention Facility.* Oklahoma City, OK: Oklahoma County Juvenile Justice Detention Center.

Kramer, Candice. 2001. *Success in Distance Learning.* New York: Delmar Publishers.

Lockwood, Fred and Anne Gooley. 2001. *Innovations in Open and Distance Learning: Successful Development of Online and Web-Based Learning.* New York: Stylus Publishing.

Lunsford, Andrea and Robert Connors. 1992. *The St. Martin's Handbook.* 2d ed. (annotated instructor's edition). New York: St. Martin's.

Morgan, A. 1991. *Research into Student Learning in Distance Education.* Victoria, AU: University of South Australia Press.

Pearce, Catherine Owens, ed. 1958. *A Scientist of Two Worlds: Louis Agassiz.* Philadelphia, PA: Lippincott.

Philliber, Susan G., Mary R. Schwab and G. Sam Sloss. 1980. *Social Research.* Itasca, IL: F. E. Peacock.

Picciano, Anthony G. 2001. *Distance Learning: Making Connections across Virtual Space and Time.* Upper Saddle River, NJ: Prentice Hall.

Report by the President's Commission on Law Enforcement and Administration of Justice. 1967. (February). Washington, D.C.: Government Printing Office.

Social Sciences Index. Annual. New York: H. W. Wilson.

U.S. Bureau of the Census. 1988. *County and City Data Book.* Washington, D.C.: Government Printing Office.

U.S. Bureau of the Census, 1971. *Historical Statistics of the United States: Colonial Times to 1970.* 1971. Washington, D.C.: Government Printing Office.

U.S. Department of Education. 1998. "Fall Enrollment Surveys." *Integrated Postsecondary Education Data System* (IPEDS). Washington, D.C.: National Center for Education Statistics.

U.S. Superintendent of Documents. Annual. *Monthly Catalog of United States Government Publications.* Washington, D.C.: Government Printing Office.

Young, Jeffrey R. 2000. "Distance and Classroom Education Seen as Equally Effective." *Chronicle of Higher Education,* February 18, pp. 6–10.

Index

A

Abstract (of paper):
    description of, 50
    example of, 52
Abstracts (databases), 115
Agassiz, Louis, 8
American Sociological Association Style Guide:
    block quote format, 41
    classic text, 66, 75
    corporate author citation, 69-70
    direct quotation, 64-65
    electronic source citation, 80
    executive department document citation, 68
    gender bias, avoiding, 21
    legal reference citation, 69
    margins, 49
    outlining, 17-18
    public document citation, 67
    references, 70-82, 83
    source citation, 65-66
    table of contents, 55
    text citations, 62-70
    use of, 62
American Sociological Review, 62
Analytical book review, 139
Analytical case study, 146, 150
Antecedent, 39
APA style. *See Publication Manual of the American*
    Psychological Association
Apostrophes, 29-30
Appendixes:
    citing, 65, 86
    description of, 59-60
    in policy analysis paper, 171
Archival sources:
    citing, 66
    reference listing, 76
Article critique:
    choosing article, 132

overview of, 129, 132
    sample, 134-137
    writing, 132-134
Article format, 59, 70-71
ASA Style Guide. See American Sociological Association
    Style Guide
Assignment:
    article critique, 129, 132-134, 135
    book review, 137-140
    policy analysis research paper, 161-171
    policy analysis research proposal, 154-160
    reaction paper, 127-129, 130-131
    understanding, 102
Atkinson, G. B. J., 145
Attitudes toward policy, 170-171
Audience:
    correctness and, 27
    defining, 13, 102
    jargon and, 20
    outlining for, 17-18
Author-date system (APA style), 83
Author name:
    reference listing, 71-72, 73-74, 89-92
    text citations, 62-64, 82, 83-85

B

Background reading, 105-106
Becker, Howard S., 145
Becker, Ronald F., 115
Benefit/cost analysis, 169-171
Bias, 12-13
Bias-free language, 21-22
Bibliography:
    working type, developing, 107
    See also Reference listing
Bills and resolutions:
    citing, 67-68, 88
    reference listing, 77, 94
Biographical research, 103

Book review:
elements of, 138-139
format and length of, 140
objectives of, 137-138
types of, 139-140
Book Review Digest, 139
Book Review Index, 139
Books:
evaluating, 116
in reference listing, 71-75, 89-92
See also Book review
Bouma, Gary D., 145
Brainstorming, 15
Bray, R., 114
Brundage, D., 124
Budget in research proposal, 160
Bureau of Justice Statistics Web site, 119
Business, public vs. private, 149-150

C

Capitalization:
rules for, 30-31
of titles in reference list, 71, 89
Carr, Sarah, 120
Case study:
description of, 144-145
limitations of, 145-146
parts of, 148-149
in research, 145
service agency type, 146-148
text of, 149-150
types of, 146
CD-ROM source, 82, 97
Chapter, citing, 65, 86
Chicago Manual of Style (CMS):
bill and resolution citation, 67-68
encyclopedia and reference book article citation, 74
government publication citation, 69, 70
interview citation, 70
newspaper citation, 66-67
personal communication citation, 80
United States Constitution citation, 68
use of, 62
Classic text:
citing, 66, 87
reference listing, 75, 92
Class notes, rewriting, 5
Clichés, 20-21
Coherence, 22-23
Cohort research, 103
Collier's Encyclopedia, 106
Colons, 30, 31-32
Commager, Henry S., 4, 6
Commas:
comma splice, 32-33
in compound sentence, 33-34

quotation marks and, 42
with restrictive and nonrestrictive elements, 34
in series, 35
Commission on Law Enforcement and Administration of Justice, 152-154
Communication:
audience and, 13
See also Personal communication
Comparative research, 103
Compensating variation, 170-171
Compound sentence, 33-34
Conclusion:
in reaction paper, 129
in social service agency case study, 150
Congressional Record, 67-68, 77
Congressional Research Service, 161
Connors, Robert, 5
Consistency, 28
Content, repetition of, 23
Content analysis, 104
Content reference work, 114
Context:
capitalization and, 31
of case study, 149-150
correctness and, 27
of writing, 8
Control of paper, 101-104
Copy:
of manuscript, keeping, 49
of source material, 109
Corporate author, citing, 69-70
Correctness, levels of, 27
Cost/benefit analysis, 169-171
Court decisions. *See* Legal references
Creativity and formal structure, 18
Criteria for evaluation of policy, 167-168
Critique. *See* Article critique

D

Dangling modifier, 35-36
Database, checking, 109, 115-116
Deep approach to distance learning, 124
Defining:
audience, 13
purpose, 11-13, 102
Department of Justice Web site, 117-119
Depth of focus in case study, 145
Descriptive language, 21
Developing thesis, 10-11, 106-107
Didactic case study, 146
Direct observation, 104
Direct quotation:
ASA Style Guide, 64-65
citing, 85-86, 111
crediting, 61
quotation marks and, 40-42

Direct quotation (*continued*):
    using, 111-112
Distance learning:
    criminal justice courses and resources, 123
    overview of, 120-122
    studying for, 123-124
Doyle, Michael, 132-133
Drafting stage:
    language choices, 19-22
    rough draft, 18-19

E

EBSCOHOST, 109
Economic environment, 166
Editing stage, 23-25
Edwards, Richard H., 106
Effectiveness of writing, 26-27
Efficiency ratio, 168
Electronic source, referencing, 80-82, 96-97
Elucidation in book review, 138-139
E-mail document in reference listing, 81, 97
Encyclopedia article:
    background reading and, 106
    referencing, 74, 92
Encyclopedia Britanica, 106
Encyclopedia of Social Work, 106
Enthusiasm, 6-7
Enticement in book review, 138
Environment:
    of case study, 149-150
    of policy, 166-167
ERIC, 115
Estimating costs and benefits of policy, 168-171
Ethical use of source material, 111-113
Ethnographic case study, 146
Ethnography, definition of, 146
Evaluating:
    article, 133-134, 137
    book in book review, 139
    book review, 139-140
    effectiveness and efficiency of policy, 167-168
    sources, 108-109, 116
Evidence, critiquing, 133, 136
Examination in book review, 138
Executive department document:
    citing, 68
    reference listing, 78
Executive summary:
    description of, 51-54
    policy analysis paper, 162-164
Experimentation, 104
Expository writing, goal of, 11-12

F

Facts:
    of case study, 149

of policy analysis paper, 162
Feedback, obtaining, 110-111
Field study, 146
Finding aid, 114
FIRSTSEARCH, 109
Flexibility, maintaining, 16
Fonts, 49
Formality, level of, 19-20
Format:
    abstract, 50, 52
    appendixes, 59-60
    book review, 140
    executive summary, 51-54
    headings and subheadings, 57-58
    illustrations and figures, 59
    lists of tables and figures, 56-57
    margins, 49
    outline summary, 54-55
    overview of, 48-49
    pagination, 49-50
    policy analysis paper, 162-164
    reference listing, 59
    table of contents, 55-56
    tables, 58
    text, 57
    title page, 50, 51
Forster, E. M., 3
Fragment. *See* Sentence fragment
Freewriting, 14
Fused sentences, 37-38

G

Geer, Blanche, 145
Gender-neutral language, 21-22
Goal of policy, 167
Government publication, citing, 69, 70
Grammar, definition of, 28
Group as author:
    citing, 85
    reference listing, 90
Guide to literature, 114
Guide to Reference Books (Bray), 114

H

Hamilton, Nancy K., 135-140
Handbook, 114-115
Hara, Noriko, 122
Hartwell, Patrick, 28
Headings and subheadings, 57-58
Historical research, 103
Historical Statistics of the United States, 115
Houck, Scott M., 130-131
Hughes, Everett C., 145
Hunter, David E., 146
Hypothesis and case study, 145

I

Illustrations and figures, 59
Independent clause, 32, 33
Indexes, 115-116
Informative writing, goal of, 11-12
Internet:
    mailing list, 119
    news group, 120
    See also Web sites
Interpretation of case study, 146
Interview:
    citing, 70
    conducting, 109-110
    reference listing, 79
    service agency case study, 148
Invention:
    asking questions, 16
    brainstorming, 15
    definition of, 13
    freewriting, 14
    maintaining flexibility, 16
Irony, 3-4
Isaac, Stephen, 145
I-syndrome, 23
Italics, use of, 83, 89
Its vs. it's, 38

J

Jargon, 20
Johnson, William A., Jr., 52, 53-54
Journal:
    articles cited in reference listing, 75-76, 81, 93, 96
    preparing paper for submission to, 48-49
Justice Quarterly format, 48-49

K

Keane, R., 124
Kennedy, John F, 5-6, 37
King, Rob, 122

L

Language choices:
    bias-free and gender-neutral, 21-22
    clichés, 20-21
    descriptive language, 21
    formality, level of, 19-20
    jargon, 20
Language competence, 26-27
Laws and statutes:
    citing, 68, 88
    reference listing, 78, 95
Learning by writing, 3, 5
Legal reference:
    citing, 69, 88
    reference listing, 78-79, 95

Legislative hearings:
    citing, 88
    referencing, 94
Levels of public policy, 165-166
Library research, 114-116
Lincoln, Abraham, 36
Lists of tables and figures, 56-57
Lunsford, Andrea, 5

M

Mackneson, R., 124
Magazine article in reference listing, 76, 93
Mailing list, 119
Maki, Ruth S., 120
Maki, William S., 120
Margins, 49
Meadows, Robert, 133-134
Measurements for evaluation of policy, 167
Memory, writing from:
    brainstorming, 15
    freewriting, 14
    invention, 13
Methods:
    critiquing, 133, 134, 136
    describing in research proposal, 159
Michael, William B., 145
Miscues, 24
Modifier, dangling, 35-36
Morgan, A., 124
Multiple sources, citing, 65, 86

N

Narrowing topic, 9-10
News group, 120
Newspaper:
    citing, 66-67, 87
    reference listing, 76, 81, 94
Nonrestrictive element, 34
Note cards, organizing, 110
Note taking, 148

O

Objectives of policy, 167
On-line citation system, 80-82, 96
On-line learning. *See* Distance learning
Online Learning Web site, 121
Options section in policy analysis paper, 169
Organizing writing, 16-18
Outline:
    formal outline pattern, 18
    policy analysis paper, 164-165
    for reader, 17-18
    research process and, 110
    for self, 17
Outline summary, 54-55

P

Page numbers in text citation, 63, 83
Pagination, 49-50
Paradox, 3-4
Parallelism, 36-37
Paraphrasing, 112, 113
Passive-voice verb, 36
Patterns of crime research, 103
Pearce, Catherine Owens, 8
Period and quotation marks, 42
Periodical, referencing, 75-76, 92-94
Personal communication:
     citing, 86-87
     reference listing, 79-80, 94
Persuasive writing, expository writing compared
          to, 11-12
Philliber, Susan G., 144, 145, 146
Photocopying source material, 109
Physical environment, 166
Plagiarism, 112-113
Planning stage:
     audience, defining, 13
     invention strategies, 13-16
     outlining, 16-18
     overview of, 8
     thesis, finding, 10-13
     topic, narrowing, 9-10
     topic, selecting, 8-9
Policy analysis:
     description of, 152
     example of, 152-154
     public officials and, 161
     research proposal for, 154-160
Policy analysis paper:
     alternatives section, 169
     benefit/cost analysis section,
          169-171
     contents of, 162-164
     description of, 161
     description section, 165-166
     effectiveness section, 166-167
     efficiency section, 168
     environment section, 166
     presentation of, 171
     purpose of, 161
     See also Policy analysis
Policy deficiency, 156
Policy problem, 156
Political environment, 166
Possessive:
     forming with apostrophe, 29-30
     forming with s, 38
Practicing writing, 6, 28-29
Presidential Commissions, 152
Primary research, 103
Privacy of source, protecting, 149

Process of writing:
     drafting stage, 18-22
     editing stage, 23-25
     overview of, 7-8
     planning stage, 8-18
     proofreading stage, 25
     revising stage, 22-23
Product of research, 160
Pronoun errors, 38-40
Proofreading stage, 25
Proper nouns, 30-31
Proposal. *See* Research proposal
Psyclit, 115
*Publication Manual of the American Psychological Asso-
          ciation:*
     author citation, 83, 84-85
     author-date system, 83
     block quote format, 41
     classic text, 87, 92
     direct quotation, 85-86
     outlining, 17-18
     public document citation, 87-88
     reference listing, 89-98
     text citations, 83-88
     use of, 62
Public document:
     citing, 67, 87-88
     reference listing, 77-79, 94-95
Public officials and policy analysis, 161
Public vs. private agency, 149-150
Purpose of writing, defining, 11-13, 102

Q

Quality control, 159
Quayle, Dan, 27
Questions, asking:
     invention and, 16
     topic selection and, 8
Quotation marks, 40-42

R

Reaction paper:
     concluding, 129
     issue, defining, 128
     overview of, 127
     position, stating and defending, 129
     sample of, 130-131
     selection of statement, explaining, 128
     statement, selecting, 127-128
Reader. *See* Audience
Reader's Guide to Periodical Literature, 115
Records research, 104
Reference book article, referencing, 74, 91
Reference listing:
     APA style, 89-98
     article format, 70-71

ASA Style Guide, 70-82, 83
    books, 71-75, 89-92
    electronic source, 80-82, 96-97
    format for, 59
    indentation style, 89
    interview, 79
    periodical, 75-76, 92-94
    public document, 77-79, 94-95
    unpublished source, 80
Reflective book review, 142
Repetition, avoiding unnecessary, 23
Report by the President's Commission on Law Enforce-
       ment and Administration of Justice, A, 152-154
Reprint:
    citing, 65, 86
    reference listing, 74, 92
Requesting information by mail, 107-108
Research:
    case study in, 145
    published work on topic, 11
    statement of need, 156
    types of, 103-104
    See also Methods; Research proposal
Research process:
    background reading, 105-106
    database, checking, 109
    feedback, obtaining, 110-111
    interview, conducting, 109-110
    library information, 114-116
    outline, drafting, 110
    overview of, 101
    rough draft, 110
    schedule for, 104-105
    thesis, developing, 106-107, 110
    topic, narrowing, 106-107
    working bibliography, developing, 107
    writing for needed information, 107-108
    written sources, evaluating, 108-109
    See also Research
Research proposal:
    comparative evaluation of policy alternatives, 158
    content of, 155-156
    description of, 154
    evaluation of current policy, 158
    methodology, 159
    policy problem and policy deficiency, 156
    problem clarification, 157-158
    product description, 160
    purpose of, 154-155
    quality control, 159
    recommendations, 158
    resources needed, 160
    statement of imperatives, 156-157
    statement of objectives, 157
Resources:
    distance learning, 123

research proposal, 160
Restrictive element, 34
Results, keeping in perspective, 104
Retrieval date for on-line source, 81
Rettig, Richard P., 52, 53-54
Revising stage, 22-23
Rewriting class notes, 5
Rollins, 40
Roosevelt, Franklin, 4
Rough draft, 18-19, 110
Run-on sentence, 37-38
Russell, Thomas L., 120

S

Schedule:
    for research, 104-105
    for research project, 160
    for writing, 19
Schwab, Mary R., 144, 145, 146
Scott, 41
Secondary research, 103
Secondary source:
    citing, 65
    reference listing, 76
Selecting topic, 8-9
Self-confidence, 7, 28
Self-discovery, 12-13
Semicolons, 42-43
Sentence:
    compound, 33-34
    fused or run-on, 37-38
    thesis type, 10, 12
Sentence fragment, 26, 43-44
Service agency case study, 146-148
Shift in person, 40
Singer, Isaac Bashevis, 5
Sloss, G. Sam, 144, 145, 146
Social environment, 166
Social Sciences Index, 115, 116
Sociofile indexes, 115
Source material:
    citing, 113
    documentation style, 61-62
    documenting, importance of, 61
    ethical use of, 111-113
    evaluating, 108-109
    photocopying and downloading,
        109
    See also Reference listing
Specific Evidence and Expert Testimony Handbook
        (Becker), 115
Spelling:
    commonly confused words, 45
    commonly misspelled words, 45-47
    computer spell checker, 44
Statement of need, 156

State University of New York at Albany Department of
  Criminal Justice Web site, 117
Statistical handbook, 114-115
Statistical yearbook, 115
Steward, Gary A., 52, 53-54
Strauss, Anselm L., 145
Style guide. *See American Sociological Association Style
  Guide; Chicago Manual of Style (CMS); Pub-
  lication Manual of the American Psychologi-
  cal Association*
Subheadings, 57-58
Subject handbook, 115
Subject yearbook, 115
Submission to journal, preparing paper for, 48-49
Subordinate clause, 26
Surface approach to distance learning, 124
Surface errors, 27
Survey research, 104, 109

T

Table of contents, 55-56
Tables:
  citing, 65, 86
  format for, 58
Text:
  of case study, 149-150
  format for, 57
  of policy analysis paper, 162, 164-165
  of research proposal, 155-157
Text citations:
  APA style, 83-88
  ASA Style Guide, 62-70
Thesis:
  article critique, 132-133, 134
  developing, 10-11, 106-107
  drafting, 110
  finding, 10
Thesis sentence, 10, 12
Time, allowing enough:
  for research, 105
  for revision, 22
  for rough draft, 19
Title page, 50, 51
Tone, 19-20
Topic:
  assignment and, 102
  interest in, 6-7, 8
  narrowing, 9-10, 106-107
  selecting, 8-9
  service agency case study,
    146-148, 150
  See also Thesis

Transitional elements, 5, 22-23
Type size, 49

U

United States Constitution:
  citing, 68
  reference listing, 78
United States Government Periodical Index, 115-116
Unity, 22
Unpublished source:
  citing, 86
  reference listing, 80
  See also Electronic source, referencing; Personal
    communication
Untranslated book, reference for, 72-73, 91

V

Vague pronoun reference, 38-39

W

Web sites:
  Bureau of Justice Statistics, 119
  Department of Justice, 117-119
  distance learning, 123
  Online Learning, 121
  State University of New York at Albany
    Department of Criminal Justice, 117
  Web U.S. Universities, by State, 123
  Western Governors University, 123
  World Campus 101, 120-121
Web U.S. Universities, by State Web site, 123
Western Governors University Web site, 123
Whitten, Phillip, 146
Wording, repetition of, 23
Working bibliography:
  developing, 107
  evaluating sources in, 108-109
Working thesis, 10, 106-107
World Campus 101 web site, 120-121
Writing:
  difficulty of, 5-7
  importance of, 4-5
  irony of, 3-4
  learning by, 3, 5
  practicing, 6, 28-29
  to request needed information, 107-108
  self-confidence in, 7
Writing context, 8

Y

Yearbooks, 115
Young, Jeffrey, R., 120